Design for Environmental Conservation

Gabriela Nuri Baron
Editor

Design for Environmental Conservation

A Toolkit that Enables Sustainable, Participatory and Distributed Innovation

Springer

Editor
Gabriela Nuri Baron
Design
University of Auckland - Waipapa Taumata Rau
Auckland, New Zealand

ISBN 978-981-96-9573-7 ISBN 978-981-96-9574-4 (eBook)
https://doi.org/10.1007/978-981-96-9574-4

This Springer imprint is published by the registered company Springer Nature Singapore Pte Ltd.
The registered company address is: 152 Beach Road, #21-01/04 Gateway East, Singapore 189721, Singapore

To Serena and Sofia, and all the ones yet to come…

Acknowledgments

To my daughters, to whom I dedicate this book, you are my guiding stars. My greatest hope is that your children, and theirs, inherit an Earth worth living in—a thriving playground for their adventures.

To my partner, Gabriel, whose unwavering love and support have been a constant source of strength in both my career and personal journey. This research has been more than an academic pursuit; it has been a journey of personal growth, and I am grateful to have shared it with you.

To my parents, who raised me immersed in nature, encouraged me to become a certified scuba diver, and took me to awe-inspiring places around the world. Your love for exploration and the natural world shaped my path.

To my sister, Dr. Maria Baron Palamar, who first introduced me to the world of environmental conservation and the challenges it faces. Your passion and dedication continue to inspire me.

This work is also a ripple effect of the generosity of Dr. Carlo Vezzoli, whose mentorship during my Master's studies profoundly shaped my career. His commitment to sharing open-access sustainability tools with the world led to the creation of the global LeNS project (Learning Network on Sustainability), a legacy that continues to inspire change.

I am also deeply grateful to my colleague Ayla Hoeta, who has patiently guided me in understanding the profound wisdom of Indigenous worldviews, particularly in Aotearoa New Zealand. Through her teachings, I am learning to embed strength, humility, and care into every project.

This work has been partially funded by the University of Auckland, Waipapa Taumata Rau.

Finally, I want to acknowledge that this work would not exist without the wisdom of those who came before us, the communities who share their knowledge, stories, and insights, the participants of my workshops, my students, and the natural systems that continue to inspire me. I am deeply grateful to the many contributors, collaborators, participants, and readers who bring these ideas to life. Let this be an ongoing, pluriversal dialogue, a shared commitment to change, and a collective effort to restore our ecosystems, working alongside Mother Earth.

Competing Interests The author has no competing interests to declare that are relevant to the content of this manuscript.

Contents

About the Authors

Gabriela Nuri Baron is an educator, researcher, and strategy consultant specializing in participatory methodologies for sustainability, conservation, and social innovation. With over 18 years of experience, she has developed a reputation as a passionate, adaptive, and strategic problem-solver. She currently serves as a Senior Lecturer and Researcher at the University of Auckland.

Her career has spanned multiple creative fields, always centered on the intersection of people and the environment. Having worked and studied across various countries and continents, she values learning through experience and finds inspiration in cultural and ecological diversity. She strongly believes in co-creation and collaboration as essential drivers of resilient innovation, and partners with Indigenous and local communities in her work, adopting a decolonial approach.

Gabriela holds a degree in Industrial Design from Argentina, a specialized Master's from Politécnico di Milano, and a PhD in Environmental Engineering. Beyond academia, she is a dedicated mother of two and a certified DiveMaster, deeply connected to the ocean and the natural world.

Maria B. Palamar is a wildlife veterinarian and conservation biologist dedicated to fostering leadership and systems change in environmental organizations. Originally from Argentina, she has spent the last 20 years in the United States, now residing in the Appalachian Mountains, the land of the Cherokee.

Maria co-founded Resolve Conservation in 2018 to help conservation and environmental organizations transform their culture, build leadership capacity, and address complex challenges through an equitable and just lens. She specializes in adaptive leadership, systems thinking, group dynamics, and trauma-informed practices, guiding organizations to become more inclusive and representative of the communities they serve. Her expertise extends to designing and facilitating capacity-building programs for state, federal, private, and nonprofit organizations.

Maria is a certified Case in Point teacher (Kansas Leadership Program) and holds certifications in Adaptive Leadership (Harvard Kennedy School) and Immunity to Change facilitation (Minds at Work, Boston). She is also a certified somatic practitioner, intuitive coach, and yoga teacher, bringing a holistic approach to her work. Maria loves spending time in nature, walking barefoot, writing, eating candy, and drinking good wine.

Ayla Hoeta is a maama of three, an educator, and an advocate for Kaupapa Maaori-centered design and innovation. With over a decade of experience in community-led projects, she is dedicated to supporting whaanau aspirations and enabling intergenerational well-being for tangata whenua and te taiao.

A passionate Maramataka practitioner, Ayla shares practices to align hauora with Maramataka tupuna knowledge.

Her whakapapa connects deeply to Taamaki Makaurau, South Auckland, grounding her work in whenua, whakapapa, and whaanau. As a Lecturer and Researcher at the University of Auckland's Faculty of Engineering and Design, Ayla's mahi weaves together design, innovation, and kaupapa Maaori, bridging these disciplines across design practice, education, and whaanau-led innovation.

Ayla holds a First-Class Honours degree in Communication Studies and a Bachelor of Arts in Maaori Media from the School of Design and Creative Technologies at AUT University.

Chapter 1
Introduction

Gabriela Nuri Baron with Contributed by Maria B. Palamar

This book invites us to think differently about environmental conservation, a call to explore new ways of being, doing, and thinking, rooted in our shared identity as humans: creatures deeply connected to the land and all living things. At its core, this is a story about interconnectedness, about recognizing that solutions to our environmental crises must follow the laws of nature itself.

It is clear now that the scale and urgency of our environmental challenges demand more than scientific and technical knowledge. They require approaches that weave together technical, relational, and cultural perspectives. The dominant Western narrative, where humans are separate from nature, has failed us.

Design for Conservation (D4C) aims to offer an alternative: a participatory, decolonial, multi-species, and distributed methodology. It combines scientific knowledge with other types of wisdom, fostering collaboration and empowering communities to co-create inclusive, sustainable, and impactful ways of tackling environmental challenges.

We Are Mammals

Humans, as mammals, have an ancestral connection to the land and all the organisms that live on it. Just like all mammals, we've evolved with an acute sensitivity to body language, an instinct that allows us to perceive and mirror the emotions of those around us. For example, when we sense aggression or stress (from other humans or animals) our bodies react instinctively, readying us to fight, flee, or

G. N. Baron (✉)
Faculty of Engineering and Design, University of Auckland – Waipapa Taumata Rau, Auckland, New Zealand
e-mail: g.baron@auckland.ac.nz

G. N. Baron (ed.), *Design for Environmental Conservation*,
https://doi.org/10.1007/978-981-96-9574-4_1

freeze. These reactions are ancient survival mechanisms, happening before our minds can even rationalize what's happening. This same instinct drives our capacity for empathy. We feel joy and peace in the presence of others' happiness and suffer when we see others in pain. This ability to *connect to our pack* is part of what makes us human, but it also leaves us vulnerable to the world's suffering.

Today, as our ecosystems degrade and the natural world struggles, we are mirroring this pain. The rising rates of anxiety and depression in humans are not separate from the state of the Earth, they are reflections of it. As mammals, we are uniquely equipped to feel and respond to the critical state of our environment. Yet, our responses vary, from numbness and despair, to denial, or to a fierce determination to fight for our planet.

Past approaches to conservation have often failed due to multiple reasons: distrust in science, the disempowerment of local communities, greenwashing, *big green*, technocracy, colonialism, and extractive economic models, among others. But if we can remember that we are mammals, we can look at our environmental and social challenges in a way that works with our ancestral and evolutionary makeup, where our natural empathy and compassion can synergize our critical thinking and factual knowledge in a way that takes care of both the problem, and the people trying to solve it. We are creatures that thrive on social connection, playfulness, and creative thinking, and only when we can fully harness the power of our human nature, will we be able to tackle the most complex challenges of our time.

Design for Conservation (D4C) was created to help us do just that: utilize our full human nature to diagnose, experiment, and resolve complex issues that require more than just the understanding of the facts. When we remember we are mammals, we begin to act from a more than rational place, a place of care, where we can fully show up and embody our humanness, honoring and harnessing the rich and intricate relationships with our ecosystem.

A Complex, Interconnected Reality

Our environmental reality is a complex combination of interconnected crises that threaten the long-term viability of our life on planet Earth. Climate change is at the heart of this reality, accelerated by the unchecked increase in greenhouse gas emissions from human activities such as burning fossil fuels, deforestation, and overconsumption.

As global temperatures rise, we witness more frequent and severe extreme weather events, such as hurricanes, floods, droughts, and wildfires, which deeply damage ecosystems, infrastructure, and communities. Animals and plants cannot adapt to these rapid changes, leaving extinction and loss of biodiversity as the natural outcomes and transforming ecosystems so dramatically that both the species that inhabit them and the essential ecosystem services they provide for human communities—such as food, clean water, temperature regulation, and carbon sequestration—collapse.

Addressing this grand, global challenge requires a coordinated effort that transcends political boundaries and requires a profound paradigm shift in the ways that Western civilizations have lived for the last 100 years. While science provides essential insights into the mechanisms of climate change and ecosystem degradation, addressing these issues requires holistic approaches that integrate science into the cultural, social, and relational dynamics of conservation. It requires us to put aside our colonial narrative of *humans as separate from nature* and draw from our mammalian ancestry, and deep connection to the land.

Participatory, distributed, and decolonial methodologies offer a promising path forward. D4C emerges as a novel methodology within this field, providing actionable tools that enable communities and stakeholders to co-create solutions relevant to their specific context. Design for Conservation bridges the gap between theoretical knowledge and actionable, sustainable conservation practices by fostering adaptability, connection, collaboration, and a commitment to purpose. This approach complements rational, scientific paradigms and extends the invitation for communities to take ownership of their environmental futures, offering a hopeful, holistic, place-based approach to restoring our ecosystems and our connection to the land.

Systemic Challenges Require Holistic Approaches

Historically, we have tended to focus on technical solutions that lack the adaptability necessary to be truly impactful, not because we cannot see the magnitude of the systemic problem, but because we often lack the tools to address larger systemic issues.

Focusing on narrow ecological challenges usually results in narrow solutions that do not address the complexity of the challenge, and are either immediately unsuccessful or may present an initial improvement that cannot be sustained over time. This is in part because of our colonial understanding of how to look at problems, the time it takes to resolve them, and how to measure success… and also because it is human nature to try to remove pain and discomfort as fast as possible, even if the relief is only temporary.

An honest, non-reductionist look at the systemic nature of complex challenges is essential (Fig. 1.1).

Let's look at the main characteristics of complex challenges:

1. Systemic: Complex challenges are part of larger systems, with multiple causes and effects at various scales. Environmental issues are always interlinked with economic, social, and political systems, making isolated solution approaches insufficient.
2. Dynamic: These challenges are not static; they are constantly changing. Each environmental issue will evolve based on new developments, such as shifts in public policy, or a global pandemic.

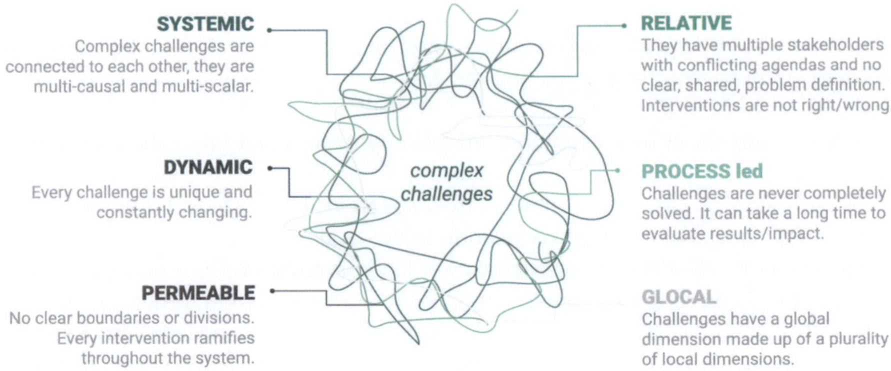

Fig. 1.1 The 6 characteristics of complex challenges. (Gabriela Baron 2024)

3. Permeable: Complex challenges lack clear boundaries, meaning interventions in one area will affect the entire system. In environmental conservation, actions taken to protect one ecosystem may have unintended consequences for others.
4. Relative: Complex challenges involve multiple stakeholders with different agendas and no single, universally agreed-upon definition of the problem. For instance, some communities will frame a conservation problem from a spiritual perspective, while others may see environmental preservation as a purely technical challenge.
5. Process-led: Solutions for complex challenges are never entirely final. They require long-term commitment, and it usually takes years, even decades, to fully understand their impacts. Because these problems are dynamic and constantly changing, conservation requires a multi-generational approach.
6. Glocal: The environmental crisis manifests in both global and local scales, each influencing and exacerbating the other. For example, climate change is a global issue but manifests in specific extreme weather events in different contexts.

Adding to this complexity, the impacts of the environmental emergency disproportionately affect vulnerable and marginalized communities, deepening existing social and economic inequalities. Indigenous communities are disproportionately affected, as their traditional ways of life are very connected to the land they have historically inhabited. These communities often have the least resources to adapt to and recover from environmental shocks, making them particularly vulnerable to the consequences of climate change.

Meanwhile, conservation efforts in Western societies have been historically managed through top-down approaches, where policies and strategies are created by central authorities or experts outside the community. These approaches, which extend the set of stakeholders beyond the most affected community (e.g., conservation professionals, scientists, politicians), usually focus on limited outcomes without considering community narratives, traditional resource uses, economic needs, and cultural sensitivities. This approach often leads to community disenfranchisement and low engagement. Without community support, the solutions are unsustainable, short-lived, and usually fail to address local needs fully.

The sustained failure of COP[1] meetings and global environmental agreements to drive meaningful change highlights the inadequacy of waiting for top-down, centralized solutions to address the climate crisis. While these efforts are essential, they are clearly insufficient. The lack of consensus, representation, political will, and enforceable commitments often stall progress when time is of the essence.

We cannot afford to wait for a unified global response. Instead, the urgency of our environmental emergency demands immediate, decentralized action that transcends political boundaries and embraces a new paradigm of bottom-up, pluriversal, place-based approaches. These should be redundant, diverse initiatives grounded in each community's unique cultural, ecological, and social contexts. While small on their own, these myriad efforts collectively coalesce into broader systemic change, creating a powerful critical mass that global treaties have failed to achieve.

When Design Thinking Meets Conservation

Design Thinking (DT) has found its way far beyond its design roots, transforming how challenges are tackled and solutions are created across fields like healthcare, education, business, and social innovation. At its core, DT is a human-centered, problem-solving approach that prioritizes empathy, creative brainstorming, learning from mistakes, and testing ideas in cycles. This structured yet flexible process has proven invaluable when addressing complex problems that require understanding people's needs and encouraging collaborative solutions.

What makes Design Thinking especially impactful is its participatory nature. Rather than treating people as passive recipients of solutions—or *consumers*—it brings stakeholders into the process, giving them an active role. By doing so, DT shifts from top-down or purely theoretical approaches to hands-on, inclusive collaboration where everyone's voice can be heard.

This led us to a big question:

How might we empower grassroots communities to design and deploy conservation strategies rooted in their unique environments and cultures, enabling them to act autonomously and effectively?

This question became the starting point for this book. As we explored how to address this challenge, we examined countless Design Thinking methods and agile tools. While many were effective and inspiring, we noticed a critical gap: most tools

[1] COP (Conference of the Parties) refers to the annual meeting of the signatories of the United Nations Framework Convention on Climate Change (UNFCCC). These conferences bring together world leaders, policymakers, scientists, and activists to negotiate and advance global climate action. Notable agreements, such as the Paris Agreement (COP21), have emerged from these summits.

didn't place Mother Earth first and foremost. Many methods still leaned on outdated mental models that failed to capture the complexities of environmental issues, often perpetuating colonialist, capitalist or polarized worldviews.

To fill this gap, we decided to hand-pick and adapt some of the existing tools, as well as develop new ones altogether. The result is a step-by-step methodology designed to guide teams through every phase of participatory design, from framing problems to implementing sustainable projects. These tools have been tailored to consider the systemic, multi-species dynamics that are central to conservation.

To be clear, we are not claiming that this is *the only or the best* approach to environmental conservation. Grounded in decolonial values, our goal was to amplify diverse ways of thinking, being, and doing. We wanted to challenge the dominant Western conservation paradigm and encourage practitioners to explore new, creative approaches that can complement it.

In the spirit of openness, we've made these tools available online—free for anyone to download, remix, and adapt. While the D4C methodology offers a roadmap for teams to start their participatory conservation journey, we encourage teams to modify, break, and reshape the method to suit their needs. The method is intentionally open, meant to inspire innovation rather than prescribe solutions.

The Design for Conservation Methodology

Design for Conservation (D4C) is a methodology designed to enable communities, conservation professionals, and other stakeholders to develop environmental solutions that are locally relevant, sustainable, and inclusive.

The D4C framework includes five main stages: Reconnect, Understand, Propose and Validate, Plan for Impact, and Deploy Sustainably. These stages offer a flexible, iterative process, allowing participants to revisit and adapt each phase based on new insights, similar to natural ecosystem evolution.

D4C proposes an *Agile-to-Deep* pathway that enables users to start with a manageable design-thinking process and gradually deepen their efforts as resources, knowledge, and community engagement grow. It also emphasizes the relational richness of forging genuine partnerships, particularly with communities that bring critical knowledge to conservation efforts..

To engage with D4C, teams can access the toolkit and resources for running workshops, understanding core steps, and exploring relational tools like the reconnection exercises that should be woven throughout the process. This toolkit enables any group—whether led by conservation experts or grassroots organizations—to facilitate a D4C workshop or process independently. Any user can download all tools in design4conservation.com and/or through the links included in this book.

How to Read this Book

Whether you're a facilitator, designer, conservationist, or someone curious about blending creativity with sustainability and environmental conservation, this book does not require previous technical knowledge, just passion and purpose.

The first few chapters of the book set the stage by introducing critical issues around conservation, decolonial approaches, participation, and facilitation. For the best experience, we recommend starting with the introductory chapters if you want to immerse yourself in the book's broader ethos. These sections provide a foundational understanding of the concepts that inform the Design for Conservation methodology, as well as the mindsets that guide this work, such as "*thinking like nature would*." If you're new to these topics, these chapters will give you the broader context you need.

As you move deeper into the book, you'll find a detailed explanation of the Design for Conservation methodology. It is organized into five phases: Reconnect, Understand, Propose and Validate, Plan for Impact, and Deploy Sustainably. Each phase includes specific tools, downloadable templates, and step-by-step guides, along with examples to help you visualize how to apply these tools in real-world scenarios.

However, this book is also designed to be flexible and non-linear. Feel free to jump directly to the tool or phase that resonates with your current needs. Whether you're seeking to plan a project, engage stakeholders, or validate ideas, you don't need to read from start to finish—browse the index and dive into what speaks to you.

- Chapter 2 explores the role of Indigenous knowledge and the importance of partnership in conservation.
- Chapter 3 discusses facilitation skills and challenges in participatory design.
- Chapter 4 introduces the mindsets that guide D4C projects.
- Chapter 5 provides an overview of the D4C methodology.
- Chapters 6, 7, 8, 9, and 10 provide a detailed explanation of each D4C phase, the tools provided, and practical examples.
- Chapter 11 outlines how to follow an agile-to-deep, iterative, and adaptive deployment of a project.
- Chapter 12 concludes with reflections and offers practical instructions for accessing and downloading the toolkit.

Whether you are familiar with design thinking methodologies or not, this book invites you to explore these new ways of thinking, being, and doing at a time when great participatory innovation is needed. Let's begin.

Chapter 2
The Need for Distributed, Decentralized, and Decolonial Environmental Conservation

Gabriela Nuri Baron with Contributed by Ayla Hoeta

Traditionally, *Western* conservation practices have relied on centralized, top-down approaches. While effective in some contexts, these methods have often excluded Indigenous and local communities, resulting in strategies that lack cultural relevance, stakeholder buy-in, and long-term sustainability. This exclusion reflects Dominant Social Paradigms,[1] rooted in colonial, patriarchal, and technocratic worldviews that prioritize economic growth and technological fixes over long-term ecological health and community resilience.

Decolonizing conservation involves shifting away from these Dominant Social Paradigms and amplifying Indigenous and local participation. Arturo Escobar's *"pluriversal thinking"* concept provides a valuable framework, emphasizing the coexistence of multiple worldviews and knowledge systems.[2] Unlike a *"universal"* approach, which seeks to globalize a single perspective, pluriversal thinking celebrates cultural diversity and proposes frameworks for the respectful, fraternal, and insightful co-existence of multiple worldviews.

In line with decolonial frameworks, conservation must also embrace a distributed and decentralized nature. Distributed approaches empower local communities to lead conservation efforts, leveraging their unique knowledge and perspectives.

[1] Dominant Social Paradigms (DSP) refer to the widely accepted beliefs, values, and assumptions that shape how societies understand and interact with the world. In the context of environmental issues, DSPs often reflect anthropocentric, growth-oriented, and market-driven perspectives that prioritize economic development over ecological sustainability.

[2] Arturo Escobar's 2020 work, *Pluriversal Politics: The Real and the Possible*, delves into the concept of the "pluriverse," advocating for the recognition and coexistence of multiple realities and knowledge systems. He critiques dominant paradigms and emphasizes the importance of embracing diverse worldviews to address global challenges.

G. N. Baron (✉)
Faculty of Engineering and Design, The University of Auckland –Waipapa Taumata Rau, Auckland, New Zealand
e-mail: g.baron@auckland.ac.nz

G. N. Baron (ed.), *Design for Environmental Conservation*,
https://doi.org/10.1007/978-981-96-9574-4_2

These decentralized systems foster resilience by allowing multiple, context-specific solutions to emerge simultaneously. This approach aligns with the principles of equity and inclusion, and increases the likelihood of long-term success by rooting solutions in the lived experiences of those most affected.

Where Multiple Worldviews Can Co-exist

The D4C methodology doesn't aim to impose one way of thinking over another. Instead, it provides a space for collaboration and mutual learning. For instance, in Aotearoa New Zealand, D4C draws inspiration from Maaori principles like *kaitiakitanga* (guardianship and stewardship towards the natural world), but it doesn't assume these practices will work the same way everywhere. Instead, the methodology encourages participants to explore how such principles might resonate within their own cultural and ecological contexts and invites them to bring forth their own worldviews into the project. This adaptability allows D4C to be used globally, blending local traditions and knowledge systems with scientific approaches.

For those unfamiliar with Indigenous knowledge systems, D4C offers tools to foster respectful engagement. These include exercises for mapping cultural values, identifying shared goals, and creating dialogue among diverse stakeholders. By prioritizing inclusion and humility, D4C creates space for all voices to shape conservation efforts collaboratively.

But how do we truly integrate Indigenous and traditional knowledge into conservation? Let's explore an example from Aotearoa New Zealand.

Honoring Kaitiakitanga: How Maaori Wisdom Shapes Aotearoa's Conservation Efforts

In Aotearoa New Zealand, Maaori have been the *kaitiaki* (guardians) of a balanced ecosystem long before colonization introduced new ideas of progress. Colonization brought not just physical pests but also mental ones: worldviews that undermined Indigenous practices and values, creating barriers to collaboration and sustainable stewardship (Smith 1999).[3]

For Maaori, the connection to *te taiao* (the environment) is profound, shaped by *koorero tuku iho* (ancestral stories) and *maatauranga Maaori* (Maaori knowledge). This resilient knowledge system emphasizes balance and reciprocity, guiding sustainable practices over generations. In *te ao Maaori* (the Maaori worldview), the environment is seen as a *tuupuna* (ancestor) with its own *wairua* (spirit) and *mauri*

[3] Smith, L. T. (1999). *Decolonizing methodologies: Research and Indigenous peoples.* Zed Books.

(life force). It's not just land—it's a living being with *mana* (spiritual power), deserving of care and respect for future generations.

Kaitiakitanga embodies this perspective, offering a holistic approach to conservation that integrates physical, spiritual, and cultural dimensions. It challenges resource-driven models and highlights the interconnectedness of all life. Practices like *rahui* (temporarily restricting resource use to allow natural recovery) demonstrate this harmony between people and the environment. Restoring these cultural connections helps heal both the land and communities affected by colonial practices.

In recent years, Aotearoa has begun recognizing *kaitiakitanga* in environmental policy. Legislation like the Resource Management Act[4] supports iwi (tribe) involvement in resource management, while local councils work alongside Maaori communities to create conservation strategies grounded in *kaitiakitanga*.

One actionable example is the integration of *Maramataka*, the Maaori environmental moon calendar, in fisheries management. *Maramataka* uses environmental signs (*tohu*) to guide seasonal activities like planting, harvesting, and fishing, aligning them with the rhythms of *te taiao* (the natural world). This knowledge, passed down through *whakapapa* (genealogy), connects communities to the environment in meaningful ways.

Regional interpretations of *Maramataka* reflect local *tohu*. For example, in Manukau, the migration of *Ngutuparore* (Godwit) aligns with local eel spawning, showing the interconnectedness of life.

If you'd like to learn more about *Maramataka*, watch this interview with Ayla Hoeta (Figs. 2.1 and 2.2).

Partnership for Conservation

Decolonizing conservation requires addressing historical injustices, elevating Indigenous knowledge systems, and fostering healing and empowerment. Indigenous approaches extend beyond conservation and should not be compartmentalized, as they embody a transformative cultural and political shift in how we relate to nature and one another. In Aotearoa New Zealand, this commitment is reflected in the Treaty of Waitangi[5] partnership structure, where Maatauranga Maaori and kaitiakitanga offer powerful frameworks for reclaiming rangatiratanga (sovereignty), asserting Indigenous rights, and restoring agency in environmental stewardship.

---

[4] The Resource Management Act 1991 (RMA) is New Zealand's primary environmental legislation, designed to promote the sustainable management of natural and physical resources. It governs land use, water, air, and coastal management, aiming to balance environmental protection with economic and social development. The RMA requires councils to regulate resource use through plans and consents, incorporating principles of Te Tiriti o Waitangi (Treaty of Waitangi) and Māori perspectives.

[5] The Treaty of Waitangi (1840) is Aotearoa New Zealand's foundational agreement between the British Crown and Māori, establishing governance while guaranteeing Māori rights to land, resources, and taonga (treasures).

Fig. 2.1 Ayla Hoeta explaining the Maramataka- the Maaori environmental/moon calendar

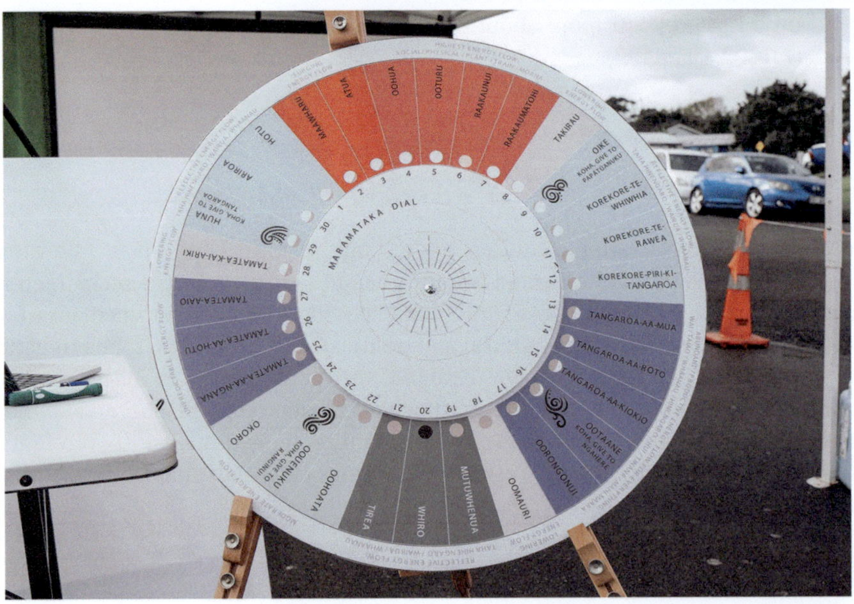

Fig. 2.2 A close-up of the Maramataka dial tool. (Designed and developed by Ayla Hoeta 2022)

For non-Maaori (tauiwi), supporting this shift means respecting and upholding kaitiakitanga, embracing maatauranga Maaori in conservation efforts, and recognizing their role as tangata tiriti (people of the treaty) and treaty partners. Through genuine partnership, we uphold Maaori as Tangata Whenua and kaitiaki, honour Te Tiriti o Waitangi according to a Maaori worldview, and work alongside Maaori to create more just, inclusive, and sustainable approaches that respect both the land and those who protect it.

Chapter 3
Holding Space for Participatory Innovation

Gabriela Nuri Baron

The concept of *the commons* refers to a set of resources shared by a community, collectively managed to ensure equitable and sustainable use. Historically, this term originated in medieval Europe, where communities stewarded shared lands under agreed-upon rules. Today, the notion of commons has expanded to include not just natural resources like forests, fisheries, and water but also digital, cultural, and knowledge-based resources.[1] Managing commons has always been complex, and the histories of colonization, extraction, and exploitation have left many communities disconnected from their common ancestral lands, culture, and resources.

Participatory methodologies are vital to addressing these historical injustices and creating sustainable conservation practices. By returning agency to local and Indigenous communities, who often hold deep ecological knowledge and a strong connection to their lands, participatory approaches ensure stewardship that respects both human and ecological needs. Indigenous worldviews, for instance, often frame sustainability as a responsibility to future generations, seeing land not as a commodity but as a trust to be honored and protected. Participatory governance models that include community consultation and decentralized decision-making can lead to conservation practices that are not only effective but also deeply rooted in the communities they aim to serve.

We don't need just a handful of conservation experts: we need millions of passionate people, each making a difference in their own way.

[1] Effective governance of commons relies on community stewardship, cooperative decision-making, and adaptive management strategies. *(International Association for the Study of the Commons, "About Commons," IASC, https://iasc-commons.org/about-commons/)*

G. N. Baron (✉)
Faculty of Engineering and Design, University of Auckland – Waipapa Taumata Rau, Auckland, New Zealand
e-mail: g.baron@auckland.ac.nz

G. N. Baron (ed.), *Design for Environmental Conservation*,
https://doi.org/10.1007/978-981-96-9574-4_3

The Design for Conservation methodology intends to embody this participatory ethos. Beyond celebrating cultural plurality and embracing adaptability, D4C relies on visual canvases and flexible formats (such as text, drawings, and performance) to engage diverse participants, making the process more accessible. Instead of prescribing solutions, D4C creates a platform for genuine collaboration, empowering communities to define their own conservation strategies.

By adopting participatory approaches, environmental projects can move beyond externally imposed solutions and foster co-creation, where local voices and ecological systems are central. This shift not only leads to more sustainable outcomes but also builds resilient partnerships, ensuring that conservation efforts thrive for generations to come.

We Need Safer Spaces for Genuine Participation

Co-design is a powerful participatory design approach that invites stakeholders—such as community members, end-users, and other key participants—into the process of addressing common challenges. These approaches help embody shared power and trust, creating space for meaningful results for all involved. However, like tackling any complex challenge, co-design comes with its own set of difficulties.

While the ideal is to engage all stakeholders throughout the process, this can be tough in practice. The most common barriers include limited resources like time, funding, and access to key stakeholders' trust. Project constraints, such as funding structures that don't cover essentials like hospitality, can limit relationship-building and discourage participation. Yet, from my experience as a facilitator, the most significant challenge is often participant availability. Many stakeholders are balancing urgent priorities in their personal or professional lives, making it difficult to commit time to co-design. Additionally, past negative experiences, such as feeling tokenized, unsafe, or seeing no tangible outcomes after hard work, can lead to understandable hesitance to re-engage in the process.

Ultimately, co-design and facilitation are not just methods; they are disciplines in their own right, requiring a unique blend of skills, knowledge, and practice that merit an entire book to fully describe. There are amazing resources out there on decolonial co-design, and I invite readers to dive deeper and deeper into them, with time, patience, and an open heart.

So what does the D4C methodology propose? Start where you are, don't wait until you feel like an expert, because the environment needs us today. Instead of focusing on delivering perfect outcomes, start small and learn from every iteration. Below I have summarised the main key aspirations for becoming a good facilitator:

1. Set Clear Expectations and Goals: Clarify objectives to align efforts and ensure participants feel their input is directed towards a shared purpose. Establish guidelines for valuing and incorporating contributions, and start the workshop with a clear plan.

2. Nurture a *Safer* Space: Create an environment where everyone feels comfortable sharing ideas without fear of judgment. Set agreements for respectful listening, open communication, and constructive feedback, especially for sensitive topics related to worldviews or lived experiences.
3. Ensure Diverse Representation: Invite participants from various backgrounds, roles, and education levels to ensure a broad understanding of issues and inclusive solutions. Make efforts to involve voices that are often marginalized, especially if the topic directly impacts those communities.
4. Reduce Power Imbalances: Be aware of group hierarchies that could hinder open conversation. Support vulnerable voices by ensuring safety in numbers and use techniques like small group discussions to equalize participation.
5. Hospitality and well-being: Make the workshop accessible by providing necessary accommodations like wheelchair-accessible spaces, interpreters, or materials in multiple formats. Virtual participation can also help include those facing logistical barriers. Hospitality is crucial, especially for Indigenous peoples, as it denotes respect. Offer nutritious food and health breaks throughout the event.
6. Cultural Competence: Recognize and respect cultural differences within the group. When working with Indigenous communities, follow appropriate customs and protocols. Understand how cultural backgrounds might influence communication styles, comfort with group activities, or decision-making preferences.
7. Adaptability: While a structured process provides certainty and manages expectations, be prepared to adapt based on participants' needs and feedback. This might mean adjusting timelines, revisiting topics, or shifting focus as new insights emerge. Flexibility allows the emergence of radical innovation.
8. Plan for different ways of expression: Use highly visual, participatory methods to encourage active involvement from everyone, understanding that expression varies. Formats like drawing, making, storytelling, and enacting can accommodate different styles and worldviews, ensuring everyone can express themselves.
9. Follow-up and Accountability: Nurturing long-term, genuine relationships of respect and care between participants is key. Communicate outcomes and next steps clearly after the workshop. Let participants know how their input influenced decisions, building trust and respect in the process.
10. Reflection and Improvement: Inclusive facilitation is an ongoing learning process. Regularly reflect on what went well and what could be improved—from both the facilitator's perspective and through participant feedback—to make future workshops more effective and welcoming.

Facilitating participatory processes is not just about guiding others: it's a journey of personal growth that challenges and deepens our abilities to listen, connect, and adapt. At its core, facilitation requires a balance between providing structure and creating space for others to step forward, trusting in the collective wisdom of the group. Developing soft skills like conflict resolution and cultural sensitivity takes time, but can grow stronger with every iteration.

Learn *from* the Ocean, Not *about* the Ocean

Finally, one of the key aspects of facilitating a D4C workshop is honouring Nature as the main stakeholder. Spending time in a natural space is a cornerstone of the Design for Conservation methodology because nature is more than a backdrop: it's an active participant in the process.

There's only so much you can learn from a publication about nature, but you can step outside and learn from the master herself. When we step outside into the beauty of the natural world, something shifts. We leave behind the walls and screens that often confine our thinking and open ourselves to the wisdom, rhythm, and presence of the environment around us.

In D4C, whenever possible, you should alternate facilitating some sessions indoors, and some outdoors. Facilitating in natural spaces gives participants a chance to ground, pause, breathe, and reconnect, not just with the land, but with themselves and each other. Nature will inspire us and teach us about resilience, cycles of growth, and the power of interdependence, offering valuable lessons that will inform the solutions we create.

Ultimately, connecting to the place we're trying to regenerate fosters understanding, respect, and a deeper sense of responsibility. When we feel the earth beneath our feet and the vitality of the ecosystems around us, conservation becomes more than a project— it allows us to move from thinking about a mindset change to truly embodying it (Fig. 3.1).

Fig. 3.1 Matriarch Circle on the west coast of Aotearoa New Zealand, exploring how other ways of being and doing can help us learn from the Ocean. (Photo by Nabulen Ignacio 2023)

Chapter 4
Think Like Nature Would

Gabriela Nuri Baron

Albert Einstein allegedly said, *"We cannot solve problems with the same thinking we used when we created them."* This idea captures the heart of what this chapter is about: transforming the way we think about environmental challenges and how we approach them. Before diving into the Design for Conservation methodology, it's essential that we first align on a shared set of values and mental models that will guide our actions. These values are not just abstract ideals; they are practical mindsets that we can embody through our actions.

Mental models or mindsets are broad overarching attitudes, beliefs, and assumptions that shape how individuals perceive and approach various aspects of life, including challenges, learning, and personal growth. While mental models help us simplify complexity, it is important to understand the biases that come with them. These models influence not only our thoughts and knowledge but also the connections and opportunities we perceive, affecting motivation, resilience, and responses to success or failure in a working team.

Unfortunately, humanity has been deeply influenced by mindsets that prioritize short-term financial gains and resource extraction over long-term ecological health. In the last 30 years, technological optimism—*Technocracy*—has become another dangerous mindset that has created complacency, as it relies on future innovations to address current environmental issues, delaying much necessary action. Technocracy also tends to externalize costs, shifting environmental and social burdens onto communities and/or future generations, ignoring the full lifecycle impacts of our actions. This *growth-at-all-costs* mentality that focuses on immediate profits has driven Western societies for too long, creating the environmental crisis we find ourselves in.

G. N. Baron (✉)
Faculty of Engineering and Design, University of Auckland – Waipapa Taumata Rau, Auckland, New Zealand
e-mail: g.baron@auckland.ac.nz

G. N. Baron (ed.), *Design for Environmental Conservation*,
https://doi.org/10.1007/978-981-96-9574-4_4

The degree of change needed to disengage from these mentalities is profound, working at the cultural level of our social systems. It is clear that we need new mindsets, but which mindsets could help us to universally re-frame our relationship with nature?

Let's start with *Empathy, Iterative Thinking*, and *Participation*, which are essential Design Thinking principles that set the groundwork for co-design.

Empathy involves deeply understanding others' perspectives and needs, ensuring solutions respond to real-world problems. In conservation, empathy extends beyond humans to include non-human species and ecosystems, recognizing that our health and survival are intricately linked with theirs. While our human perspective may limit this empathy, even the limited efforts to understand the needs of other species help bridge the gap between humans and the natural world.

Iterative Thinking is an approach that focuses on continual improvement through prototyping, testing, and refining. In D4C, this process is applied on a scalable basis, following an *Agile-to-Deep* spiral model. Teams start small, tackling problems with the resources and knowledge available, and as they learn more, they deepen their engagement. Each iteration builds confidence, encouraging teams to invest more resources, bring in experts, and adapt to new insights.

Finally, as explained in Chap. 3, **participation** is vital. Solving complex environmental issues requires diverse perspectives and collective effort. Collaborative approaches ensure that solutions are innovative, comprehensive, and deeply rooted in the communities they serve.

While these 3 principles can help us find new paths for *thinking differently*, we still need to align with how natural systems work. Below we introduce the seven Design for Conservation (D4C) mindsets that build upon these 3 principles.

The 7 Design for Conservation Mindsets

Nature is one of the greatest teachers when it comes to solving problems. By observing how ecosystems thrive—adapting, cooperating, and regenerating—we can learn invaluable lessons for tackling today's environmental challenges. These lessons have inspired the 7 Design for Conservation (D4C) Mindsets, a set of guiding attitudes that help us align our thinking and actions with natural systems.

As you explore these mindsets, you'll find they aren't rigid rules but flexible principles that adapt to different contexts at many levels. They provide a foundation for aligning human creativity with the wisdom of natural systems, re-connecting us to our true nature, enabling us to create solutions that are not only effective but truly sustainable. Let's dive into each one and explore how they can guide our facilitation styles (Fig. 4.1).

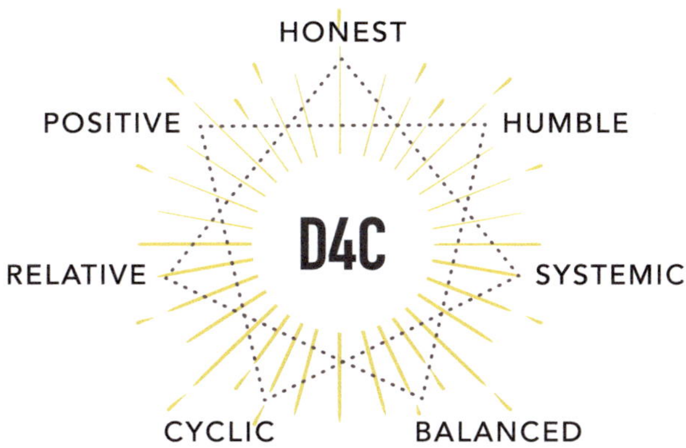

Fig. 4.1 The seven design for conservation mindsets. (Developed by Gabriela Baron, artwork by Yamila Palatnik 2021)

Honest

Any working team (and the natural realm) will benefit more from your uniqueness, your passions, and your spontaneity.

In nature, form follows function, allowing organisms to achieve their goals with minimal resources. This principle of transparency and authenticity can be applied to human interactions and teamwork. An honest mindset involves embracing your uniqueness, passions, and spontaneity, which can significantly benefit your team and the natural environment (Fig. 4.2).

In practical terms, this means:

- Embracing Your Uniqueness: Recognize and celebrate what makes you different. Your unique perspectives and ideas add value to your team and can lead to innovative solutions.
- Fostering Vulnerability and Openness: By being open about your strengths and weaknesses, you create a safe space for others to do the same. This openness builds trust and encourages honest communication.
- Building Strong, Genuine Relationships: Authentic interactions lead to deeper connections. When people feel they can trust you, they are more likely to collaborate effectively and support each other.
- Staying True to Your Core Values: Consistently aligning your actions with your beliefs ensures that you remain authentic and trustworthy. This alignment helps you navigate challenges with integrity and build lasting relationships.

Positive

Aim high and expect positive results: believe in the restorative power of nature as your ally.

Fig. 4.2 *The "Honest" mindset card.* (Developed by Gabriela Baron, artwork by Yamila Palatnik 2021)

Even the most difficult complex problems can be approached effectively with a positive mindset. Just as the natural world is a self-healing organism, any project that seeks to restore balance will find nature's inherent power as an ally (Fig. 4.3).

Positivity within a project is crucial, as it helps teams turn ideas into reality. When team members maintain a positive outlook, they are more likely to be creative, resilient, and resourceful. This mindset encourages perseverance and the belief that solutions are attainable, even in the face of adversity.

To cultivate a positive mindset:

- Harness Nature's Power: Believe in the restorative power of nature and incorporate sustainable practices into projects. Nature's resilience can support and amplify your efforts to restore balance.
- Cultivate Error-Friendliness: Encourage a culture where mistakes are seen as valuable learning opportunities, every obstacle is a chance to learn and grow. Approach every project as a journey of learning. Whether or not the immediate goals are met, the experiences gained contribute to the team's overall growth and future success.
- Be Grateful and Hopeful: Practicing gratitude and maintaining hope can boost morale and motivation. A grateful attitude helps teams appreciate their progress and strengths, while hope inspires them to keep striving toward their goals.

Fig. 4.3 *The "Positive" mindset card.* (Developed by Gabriela Baron, artwork by Yamila Palatnik 2021)

- Use Your Intent: Align your actions with positive intent. When teams work with clear, positive intentions, they are more focused and driven, enhancing their ability to achieve desired outcomes.

Humble

Trust the earth's old, wise, time-tested ways of doing things. Learn from them, try to imitate them as much as possible.

The humble mindset encourages us to trust in the time-tested wisdom of natural systems. By recognizing that nature has perfected its methods over millennia, we can learn from and find inspiration in these processes. Understanding ourselves as small, embedded parts of a vast, complex system shifts how we approach projects. Instead of assuming we can control and direct outcomes, humility prompts us to embrace a beginner's mindset, where we remain open to continuous learning rather than assuming we have all the answers (Fig. 4.4).

Humility involves acknowledging that our perspective is limited and that our human paradigms, models, and data only capture a fragment of reality. In the realm of innovation, balancing rapid knowledge acquisition with respect for slow, deep wisdom is crucial. This balance recognizes that while we can make strides in

Fig. 4.4 *The "Humble" mindset card.* (Developed by Gabriela Baron, artwork by Yamila Palatnik 2021)

knowledge and technology, we must accept that we cannot fully control natural systems. By aligning projects with natural principles and understanding the risks of rapid innovation in a space of slow wisdom, teams can track and measure both intended and unintended impacts.

To adopt a humble mindset:

- Embrace Continuous Learning: Approach each project or challenge with the eyes of a beginner, acknowledging that there is always more to learn. Stay curious and open to new insights from various sources, including nature, tried-and-tested wisdom, and emerging research.
- Align with Natural Systems: Study how ecosystems function and apply these insights to create solutions that are harmonious with nature. This approach respects the wisdom embedded in natural processes.
- Acknowledge and Accept Limitations: Recognize and accept the limitations of your knowledge and control. Understand that human models and data represent only a small part of the broader reality. Do not try to fix what is not broken.

Systemic

Understand yourself and your projects as living processes, embedded and intertwined within all of the other processes that make up a living world.

The systemic mindset is about understanding yourself and your projects as interconnected parts of a larger, dynamic system. It recognizes that nothing exists in isolation; every element within a system interacts with and influences others. This perspective acknowledges the deep importance of relationships and interactions, encouraging a holistic approach to decision-making that considers the broader context and potential impacts (Fig. 4.5).

In nature, everything functions as part of a complex web where the whole is greater than the sum of its parts. This dynamic interplay—between order, disorder, and organization—reflects the interconnectedness of all living systems. By adopting a systemic mindset, we acknowledge that humans and their projects rely on these networks of relationships and resources too.

This mindset aligns with the concept of holarchy, where systems are seen as nested layers, each one (a "*holon*") acting as an independent whole while simultaneously being part of a larger whole. For example, a single ecosystem is both self-contained and part of the global biosphere. Understanding this helps us recognize how changes in one element can ripple through the entire system.

In practical terms, this means:

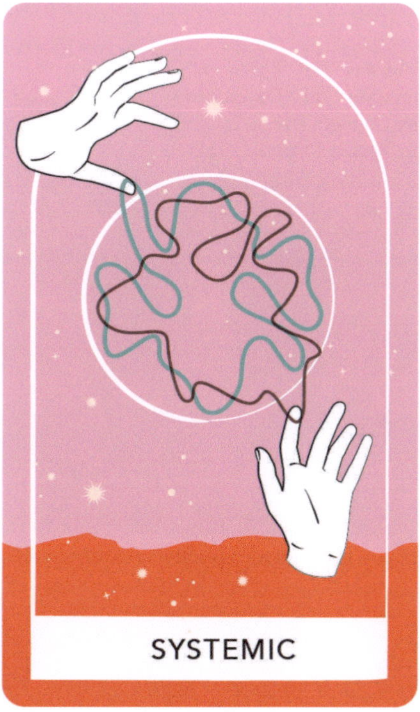

Fig. 4.5 The "Systemic" mindset card. (Developed by Gabriela Baron, artwork by Yamila Palatnik 2021)

1. Recognize Connections: Understand that every action impacts the larger network. This helps in making decisions that consider long-term and wide-ranging impacts.
2. Foster Collaboration: Building interdisciplinary teams encourages diverse perspectives, which are crucial for navigating complex, interconnected challenges. Transdisciplinary collaboration across fields often leads to sustainable and robust solutions.
3. Anticipate Ripple Effects: Plan for both intended and unintended consequences by thinking about how changes in one part of a system might affect the whole. This approach helps manage risks and improves adaptability.

Balanced

Seek balance between soft and hard approaches, they are complementary, interconnected, and interdependent in the natural world. The balanced mindset seeks to integrate polarities, in the form of feminine and masculine traits, recognizing their complementary and interconnected nature. It embraces the Yin and Yang principle, acknowledging that balance between traits like empathy and assertiveness leads to more holistic and effective solutions. By complementing qualities from both approaches, teams can achieve meaningful projects that integrate efficiency and care (Fig. 4.6).

Fig. 4.6 The "Balanced" mindset card. (Developed by Gabriela Baron, artwork by Yamila Palatnik 2021)

A balanced mindset involves integrating both *fluid* and *rigid* approaches and recognizing their complementary, interconnected, and interdependent nature. This concept can be illustrated through the *Yin and Yang* principle, representing duality balanced within an indivisible whole. Yin and Yang describe how seemingly opposite forces are actually complementary, giving rise to one another through their balanced interactions.

Yin represents feminine energies such as being, waiting, and caring, while Yang represents masculine energies such as doing, efficiency, and achieving results. Masculine traits are dominant, strong, independent, assertive, brave, disciplined, and rational. In contrast, feminine traits are emotional, collaborative, nurturing, vulnerable, caring, humble, intuitive, creative, and understanding. Think of it as a perfume bottle: Yang is the physical container, Yin is the essence it contains. Both are needed to experience the delightful scent.

For far too long, western institutions have glorified, prioritized, and funded masculine approaches, prioritizing traits like assertiveness, efficiency, and technological progress. This bias has marginalized feminine qualities, especially in fields like science and technology. The dominance of masculine approaches has led to a focus on exploitation, control, and short-term gains, often at the expense of long-term sustainability and environmental stewardship.

By sidelining feminine approaches, these institutions have created a disconnect between human activity and the natural world. The concept of *Mother Earth,* often symbolized as female, emphasizes the nurturing and life-giving qualities of nature: qualities essential for sustaining balance and health in our ecosystems.

To nurture a balanced mindset:

- Adopt an integrated Approach: Balancing feminine and masculine traits ensures a comprehensive approach to problem-solving and decision-making. Masculine traits such as assertiveness and efficiency are well complemented by feminine traits like empathy and collaboration.
- Seek Balance Within Yourself and Your Team: Embracing both sets of traits helps individuals become more self-aware, adaptable, and emotionally intelligent. Masculine traits such as independence and assertiveness help individuals take initiative and lead, while feminine traits like communication and empathy build trust and cooperation. This synergy strengthens team dynamics and improves overall performance.
- Remember the Feminine Aspect of Earth: Acknowledge and honor the Earth's feminine qualities, integrating nurturing and caring practices into your work.

Cyclic

When we think in cycles, nothing ever is lost, everything is transformed, just like in the natural realm.

The cyclic mindset views processes as ongoing cycles rather than linear progressions. It reflects the rhythms of nature, where growth, rest, and transformation are continuous and interconnected. Just as ecosystems follow seasonal patterns or life cycles, this mindset encourages embracing iterative processes that prioritize balance (Fig. 4.7).

Rather than chasing relentless growth, the cyclic mindset focuses on thriving within natural limits. This means integrating regenerative and circular models into projects from the start.

The Design for Conservation methodology embodies this mindset by encouraging iterative development through cycles of reflection, feedback, and refinement. Moving from *"agile to deep,"* this approach allows projects and teams to continuously adapt, improve, and build on prior insights. This iterative process mirrors personal growth as well, where each step forward is informed by what has come before.

To adopt a cyclic mindset:

- Approach problems as ongoing challenges: Recognize that solutions are rarely final; they evolve as contexts and conditions change.
- Think in processes and spirals: Embrace interconnected stages, where each cycleinforms the next, allowing for holistic and adaptive outcomes.
- Prioritize thriveability over endless growth: focus on creating systems that flourish within natural boundaries, ensuring sustainability and resilience over time.
- Let it die: Some projects have a short life, and that makes them perfect. In nature, nothing lasts forever, but everything transforms.

Fig. 4.7 The "Cyclic" mindset card. (Developed by Gabriela Baron, artwork by Yamila Palatnik 2021)

Relative

Be open to new perspectives within yourself and from other species. Unlearn. Challenge your assumptions. Relearn. Repeat.

The relative mindset embraces the coexistence of a pluriverse of worldviews. It encourages openness to diverse voices, continuous questioning of assumptions, and acceptance of ambiguity. By understanding that solutions are context-dependent, this mindset fosters understanding, flexibility, and respect (Fig. 4.8).

In the D4C approach, participants are encouraged to consider multiple perspectives at once, becoming comfortable with uncertainty and ambiguity. It's about being critical of existing paradigms and approaching the world with fresh eyes. In both biology and design, what works in one context may not apply in another. We must be willing to unlearn and adapt our thinking continuously. Ultimately, the relative mindset teaches us that understanding is fluid, and solutions evolve as we broaden our perspective and adapt to new contexts.

To adopt a relative mindset:

- Seek Diverse Perspectives: Engage with people from various backgrounds, disciplines, genders, and cultures to challenge your assumptions and expand your understanding.
- Cultivate Curiosity: Approach problems as opportunities for exploration. Ask questions, experiment, and remain open to new insights.

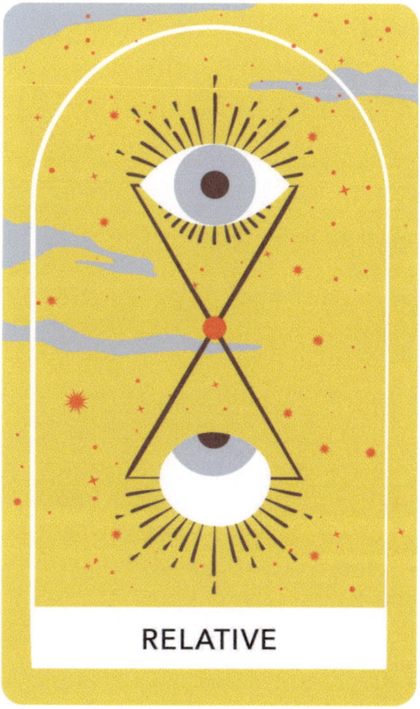

Fig. 4.8 The "Relative" mindset card. (Developed by Gabriela Baron, artwork by Yamila Palatnik 2021)

- Embrace Ambiguity: Accept that complex issues often don't have clear-cut answers. Instead of looking for a single solution, be open to multiple possibilities and be willing to adjust your approach as new ideas emerge.

In conclusion, the D4C methodology isn't about applying the same conservative approaches, but about shifting our mindset to embrace new perspectives and to design with an awareness of the interconnections between people, the environment, and all systems. These mindsets are rooted in the belief that we must work with nature, not against it.

Trust the Process

Nature is inherently self-healing: a resilient system that has evolved over billions of years to regenerate and adapt. Forests regrow after fires, rivers purify themselves over time, and ecosystems find balance even after human disruptions. This self-repair isn't just remarkable: it's instructive. It reminds us that when we create space for recovery and align our actions with natural processes, we don't need to force solutions. Instead, we can support the systems already at work.

Chapter 5
Overview of the Design for Conservation Methodology

Gabriela Nuri Baron

We have established that the environmental challenges we face demand new ways of thinking, being, and doing. In the previous chapter, we have presented new ways of *thinking* through the seven mindsets of D4C, now, let's move on to the tools that will facilitate your *doing*. This chapter takes you through the design for Conservation methodology step by step. We'll explain each phase, introduce the tools within them, and share examples of how they've been applied.

The D4C methodology comprises five interconnected stages: (Re)Connect, Understand, Propose and Validate, Plan for Impact, and Deploy Sustainably. These stages aren't linear but organized in a spiral, with the (Re)Connect stage at the center, acting as a guiding star to keep your efforts purposeful and grounded (Fig. 5.1). There are 22 tools that guide facilitators and participants through each stage. These tools are designed to be adaptable, offering clear instructions on their goals, steps, required time, and materials, but also leaving space for customization (Fig. 5.2).

Whether you're working on a grassroots community project or scaling a larger initiative, these tools help you stay true to your goals while navigating the complexity of conservation work.

G. N. Baron (✉)
Faculty of Engineering and Design, University of Auckland – Waipapa Taumata Rau, Auckland, New Zealand
e-mail: g.baron@auckland.ac.nz

G. N. Baron (ed.), *Design for Environmental Conservation*,
https://doi.org/10.1007/978-981-96-9574-4_5

Fig. 5.1 The Design for Conservation methodology comprising 5 stages organized in a spiral. (Developed by Gabriela Baron, artwork by Yamila Palatnik 2021)

(Re)Connect

The (Re)Connect stage is the foundation of the D4C process. It's where participants ground themselves, foster empathy, and build a shared sense of purpose. By focusing on connection—to oneself, the community, and the environment—this stage sets the tone for authentic, purpose-led problem-solving.

Tools:

- Thriving Earth Meditation: Cultivate presence, connection, and purpose.
- Mindset Cards: Encourage deep reflection in alignment with nature.
- Botanical Empathy: Explore your plant-self in your cultural ecosystem.
- Reconnection Rituals: Create shared moments of connection and purpose.
- Reconnection Stories: Share narratives to deepen empathy and trust.
- Climate Change Game: Move, laugh, and reflect on climate change together.

Fig. 5.2 The Design for Conservation toolkit offers a set of highly visual tools that facilitate participatory innovation

Understand

This stage focuses on solving the right problem by gathering insights, exploring the problem's context, and understanding the stakeholders involved. By taking the time to listen and observe, teams can develop a well-rounded view of the challenge at hand, frame it, and make agreements.

Tools:

- Conservation Challenge Canvas: Map out the core issue, it's context, and frame the challenges to be addressed.
- Stakeholder Characterization: Identify and understand key participants and allies.
- Stories as Knowledge: Capture wisdom from community narratives to inform decision-making.
- Agile Interviews: Gather insights through genuine conversations.
- Immersive Excursion: Deepen understanding through firsthand experience and connection to place.

Propose and Validate

This stage draws inspiration from nature, combining slow, ancient wisdom with contemporary approaches to develop innovative, resilient solutions. Rapid prototyping helps teams test ideas quickly, learn from failure, and refine their approach.

Tools:

- Inspiration Lotus: Spark creative thinking through cross-inspiration.
- Idea Selection Checklist: Prioritize the best ideas in a participatory and inclusive way.
- Idea Visualization: Communicate your idea through compelling hero shots.
- Feedback and Iteration Tool: Refine solutions based on stakeholder comments.
- Service Blueprint: Orchestrate and visualize the delivery of your project.

Plan for Impact

Impact measurement is crucial for any genuine conservation project. This stage helps teams evaluate both intended and unintended effects, ensuring projects remain sustainable and effective over time.

Tools:

- Sustainability Radar: Visualize sustainability goals and innovation progress.
- Impact Planning Tool: Create actionable impact strategies.
- Unintended Impacts Canvas: Anticipate and address potential ripple effects.

Deploy Sustainably

The final stage is where projects take shape. It's about turning ideas into action while remaining open to fix emerging errors. Most projects begin small and local before scaling up or adapting to new contexts.

Tools:

- Contextual System Map: Understand the relationships and flows in your project.
- Motivations Matrix: Align stakeholders around shared goals.
- Sustainable Growth Plan: Develop strategies for long-term success.

As you move through the D4C process, you'll find that it's not about prescribing one-size-fits-all solutions. Instead, it's an invitation to use your creativity, empathy, and connection to co-create solutions that are meaningful and sustainable. Each tool in the methodology is a stepping stone, helping you explore challenges deeply while staying grounded in your values and purpose.

When Can the Design for Conservation Methodology Be Particularly Useful?

The Design for Conservation (D4C) methodology is useful for addressing conservation challenges that require interdisciplinary collaboration, participatory approaches, and systemic design solutions. Below are some examples of cases where it can be most effective:

1. When an Interdisciplinary Approach is Needed

   Conservation problems are complex and require expertise from various fields, including ecology, design, policy, and community and Indigenous knowledge. D4C facilitates cross-sector collaboration to develop holistic, sustainable solutions.

   Example: Developing a marine conservation strategy that integrates scientific research, Indigenous ocean stewardship, and sustainable economic models.
2. When Community Buy-In is Essential

   Long-term conservation success depends on community engagement and local ownership of initiatives. D4C fosters participatory design processes to ensure solutions align with people's livelihoods and are culturally meaningful.

   Example: Co-designing a sustainable fishing program with coastal communities to balance marine biodiversity protection with economic sustainability.
3. When Innovation is Needed to Overcome Barriers

   Traditional conservation models may not always be effective in addressing urgent environmental crises. D4C encourages creative problem-solving, prototyping, and adaptive strategies.

   Example: Recovering after extreme weather events in a certain location—bottom-up climate change adaptation.
4. When Applying for Funding and Demonstrating Impact

   Funders require clear impact planning and measurable outcomes. D4C provides structured methodologies for evidence-based strategies, ensuring funding applications effectively communicate conservation goals and impact measurement strategies.

   Example: A grant proposal for a regenerative tourism initiative that includes community workshops, biodiversity assessments, and a long-term impact framework.
5. When Bridging Policy and Conservation Efforts

   D4C translates conservation needs into actionable policies, helping decision-makers integrate ecological considerations into governance frameworks.

   Example: Collaborating with city councils to design urban green spaces that enhance biodiversity while meeting community needs.
6. When Scaling Conservation Efforts Across Regions

   D4C allows for adaptable conservation models that can be scaled and replicated across different ecological and socio-political contexts.

Example: Expanding a successful mangrove restoration project by tailoring it to new locations with distinct environmental and cultural factors.

These are just some examples of practical applications. Whether you're a seasoned facilitator or new to this field, D4C offers a roadmap to navigate complexity, embrace collaboration, and take meaningful action, but remember that you can always customise and adapt the method.

Let's dive deeper into each Phase and tool.

Chapter 6
Reconnect

Gabriela Nuri Baron

The (Re)Connection Phase in participatory design is central to all others, as it builds the foundation for authentic, innovative problem-solving. This phase is designed to help participants reconnect on three levels: with themselves, with their community, and with the natural world around them.

It's beautiful how this phase taps into something deep within us: the truth that we were never separate to begin with. Our connection to ourselves, our communities, and the natural world has always been there; we've simply drifted away, lost in the noise of modern life. The (Re)Connection Phase isn't about learning something new, it's about remembering.

Through practices like mindfulness, storytelling, deep listening, and play, we gently reawaken what has always been within us. This phase is not just a preliminary step but a guiding thread woven throughout the entire journey. By returning to it often, we ensure that our work remains rooted in shared trust, authentic collaboration, and a sense of belonging.

The (Re)Connect stage sets the tone for the entire project by:

- **Embodying Mindsets:** Cultivating attitudes that are positive, empathetic, committed, creative, and adaptive.

G. N. Baron (✉)
Faculty of Engineering and Design, University of Auckland – Waipapa Taumata Rau, Auckland, New Zealand
e-mail: g.baron@auckland.ac.nz

G. N. Baron (ed.), *Design for Environmental Conservation*,
https://doi.org/10.1007/978-981-96-9574-4_6

- **Anchoring Purpose:** Keeping the team grounded in shared values and collective intention.
- **Creating a Safer Space:** Fostering trust, empathy, and openness, so that every participant feels valued and inspired to contribute fully.

Reconnection is the foundation for innovation, collaboration, and meaningful action, because only when we truly feel connected can we create lasting change.

Creating a *Safer* Space to (Re) Connect

Environmental issues are complex and multi-layerd, and require interdisciplinarity: the integration of knowledge from various fields. When working across disciplines, worldviews, and lived experiences, it's inevitable that someone might feel unsafe at some point. A "safe space" doesn't mean we can completely avoid conflict or discomfort, but it does mean focusing on creating an environment where people feel respected, heard, and supported. We can't guarantee that everyone will always feel safe, but we can foster a culture of care where relationships are nurtured and conflicts are addressed with empathy and respect.

Trust is the foundation of teamwork, allowing members to share ideas and concerns without fear of judgment, fostering creativity, and the generation of bold, innovative ideas. This inclusive approach nurtures buy-in from external stakeholders and communities, which is essential for the long-term sustainability of environmental projects.

The (Re) Connection stage is designed to designate space and time for:

- Empathy and Vulnerability: Offering activities that encourage participants to understand and share the feelings of others helps build strong interpersonal connections. Encouraging participants to share personal stories and experiences fosters trust and openness.
- Mindfulness and Reflection: Starting sessions with mindfulness exercises or meditations helps participants center themselves and become fully present. Additionally, regular reflection sessions allow participants to process their thoughts and feelings, making them feel heard and understood.
- Connection to Community and Land: Activities that reconnect participants with the land and community, such as nature walks or cultural rituals, foster a sense of belonging and purpose.
- Ice-Breakers and Team-Building Activities: Playful and engaging ice-breakers help break down barriers and create a relaxed, friendly atmosphere.
- Feedback Loops: Regular check-ins and feedback sessions allow participants to voice concerns and suggestions, making them feel valued and involved in shaping the process.

Let's dive into the tools for Reconnection.

D4C Mindset Cards

Difficulty: Easy
Duration: 5–10 min per participant (plus 10 min opening and closing)
Materials: Cards (one set per group)
Template link

The D4C Mindset Cards are a bio-inspired tool to help you adopt new ways of thinking that align with how nature works, and how we, as part of nature, can connect to those principles. There are seven D4C mental models covered in the cards, each one designed to foster a genuine connection with yourself, others, and the environment. Chap. **4** of this book dives deep into the importance of mindsets, while this tool helps facilitators ground the mindset discussion and help apply it to the project at hand (Fig. **6.1**).

How They Work

This set of eight double-sided cards can be downloaded, printed, and used in a variety of ways, whether at home or in workshops. Inspired by the feel of tarot or oracle cards, they encourage participants to tap into a different kind of wisdom and reflection.

This is the most open-ended tool in the D4C toolkit. Here are a few ideas to get started, though you are very welcome to make your own approach:

Fig. 6.1 The D4C mental model cards. (Developed by Gabriela Baron, artwork by Yamila Palatnik 2021)

1. Self-Inquiry: Lay the cards face up, ask a question, and let participants select one that "calls" to them. They can then reflect on the back's description and, after some minutes, share their thoughts.
2. Team Reflection: Ask everyone to choose a card that speaks to an essential quality for team dynamics, then discuss why they made their choices.
3. Mindset Alignment: Consider each mindset and brainstorm how the project can integrate these qualities. This can be a great way to innovate or find new directions.
4. Personal Check-In: Use the cards to explore what each person feels they need most at the moment or what their challenge is within the project.

Example Use

These cards are a versatile and meaningful tool for fostering connection and reflection, adaptable to different stages of your project. In my experience facilitating with them, they're always a hit. The cards create a space for integration and thoughtful reflection, making them perfect for following up on any of the other tools. They help re-align the team around their shared values, acting as a guiding star for the group.

In situations where participants come from different hierarchical levels (e.g., managers and employees), the cards help bridge gaps by fostering vulnerability and reducing emotional distance. They also provide insight as a conflict-resolution tool, externalizing issues and enabling participants to share their concerns or past experiences in a constructive way (Fig. 6.2).

Fig. 6.2 Interdisciplinary team member using the Mindset Cards to explore their positionality and values in relation to a conservation project. (Photo by Eric Jang 2022)

Botanical Empathy

Difficulty: Easy
Duration: 15 min to complete (plus 15 min reflections)
Materials

- Template printed in A4 or A3 size—one per group
- Drawing materials (pens, pencils, markers, colors)

Template link

The Botanical Empathy tool helps facilitators create a space where participants can connect meaningfully with each other and with the environment by using plants as metaphors. Through this process, participants can explore their own identities and roles within their communities, celebrate diversity, and build genuine, respectful connections. This activity works exceptionally well in settings related to flora conservation, such as riparian planting or reforestation projects.

How It works

1. Distribute Materials: Hand out one template per person and provide drawing supplies for everyone.
2. If You Were a Plant, What Plant Would You Be?: Ask participants to think about what type of plant best represents them. They can pick a specific plant or use general traits, like "delicate" or "resilient." The goal is to express their true selves—not what's expected, but the plant they feel identified with.
3. Community Role: Encourage participants to consider what they "provide" to their community. Are they like a plant that offers shade, nourishment, beauty, or medicine? Use this as a metaphor to reflect on their contributions.
4. Challenges and Aspirations: Invite participants to use plant imagery to describe their current state and hopes for the future. Are they like a seed, full of potential? Or maybe like a tree in bloom, vibrant and thriving?
5. Environment for Thriving: Ask participants to reflect on what they need to grow and feel at their best, be it more sunlight, connection, nourishment, or space. They can use this to talk about their needs.
6. Draw Your Plant Self: Have participants sketch their "plant selves" based on their responses. Adults LOVE to draw, some people with find it challenging, but having the opportunity to express themselves in a different way, is refreshing after a long workshop session.
7. Group Sharing: Gather everyone in a circle to share. Ensure this is a safe, judgment-free zone with no phones or pictures. People can describe their plant or simply share a few keywords if time is limited. They're also welcome to display their drawings.

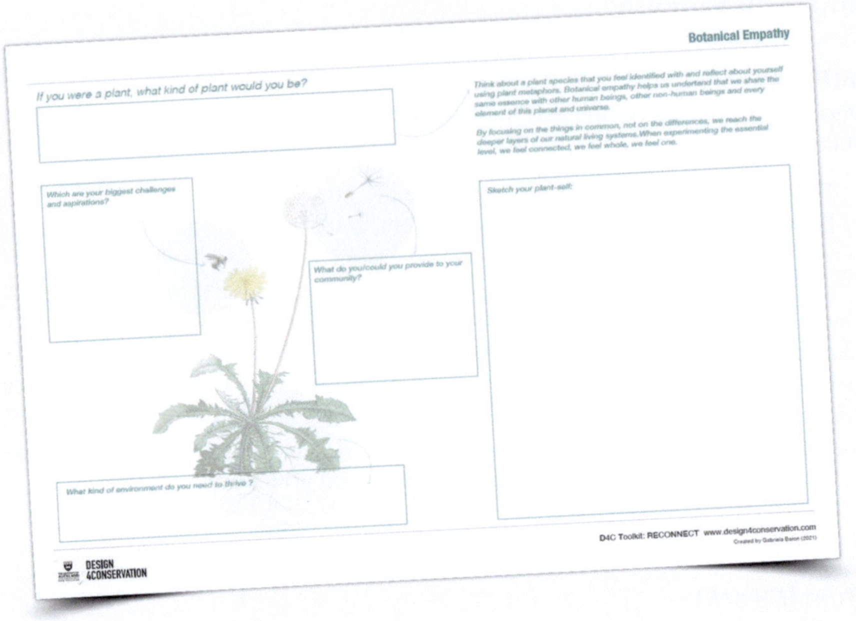

Fig. 6.3 The botanical empathy template. (Gabriela Baron 2021)

This exercise encourages self-reflection and opens the door to empathy and understanding, allowing participants to see each other's strengths, needs, and aspirations in a new light (Fig. **6.3**).

Example Use

Personally, I see myself as a dandelion, this is why I have used the dandelion image in the background of the template. This low-profile plant can grow almost anywhere, symbolizing resilience and adaptability. Dandelions are known for their ability to thrive even in the most unlikely places, like cracks in the pavement. Beyond just surviving, they offer nourishment as an edible plant and are known for their medicinal properties. When it matures, the dandelion transforms, spreading its seeds through the air, giving people a moment to pause, make a wish, and send it off into the world. This plant represents endurance, transformation, and the small but meaningful impact each of us can have.

In her book Emergent Strategy, A. M. Brown[1] describes dandelions as "*a community of healers waiting to spread*" I just love the metaphor.

[1] Brown, A. M. (2017). *Emergent Strategy: Shaping Change, Changing Worlds.* AK Press.

Fig. 6.4 This image shows the work of a participant in a workshop carried out in Huilo-Huilo, Los Lagos, Chile. He described the "siete venas" plant, also called *Plantain*. Similar to dandelion, this is a very low-profile herb that brings amazing healing qualities and has been essential for indigenous communities across the globe

While facilitating with this tool, I have seen some timid participants share their voices for the first time, opening up and sharing stories about the plants they have chosen. The most amazing thing about this exercise is that after everyone presents, you are left with a true ecosystem of plant-humans, each with their uniqueness and role, each a whole within themselves and in service to a greater whole. From tall and robust Kauri trees,2 to delicate flowers, to lichens, you will see how this ecosystem of plants reflects the complementary and interdisciplinary skills that each participant brings into the project, and they will see it too! (Fig. 6.4).

[2]The Kauri tree (*Agathis australis*) is an ancient and iconic conifer species native to Aotearoa New Zealand. Known for its towering height, massive girth, and long lifespan—sometimes exceeding 2000 years—Kauri plays a crucial role in forest ecosystems. These trees produce durable timber and resin historically used by Māori and early European settlers. Today, Kauri forests are under threat from *Phytophthora agathidicida*, a soil-borne pathogen causing Kauri dieback disease, leading to significant conservation efforts.

Thriving Earth Meditation

Difficulty: Easy
Duration: 15 min.
Materials

- Open space
- Template printed in A4 size—one copy for the facilitator OR sound player for recording
- Optional: chairs, cushions
- Script template
- Recording link—credits: Created by Gabriela Baron (2021), voice by Isla Grace (2024)

This guided meditation offers a mindful way to foster connection among participants and with the natural environment. It's designed to help people explore a shared sense of purpose and reflect on what they each want to create and nurture in the world. Through breath, physical sensations, and the Earth's energy, it offers a pause to nurture gratitude and empathy, before returning to the work at hand.

By focusing on the common threads that bind us to other humans, non-human life, and the natural elements, the meditation invites us to step beyond surface differences and acknowledge our natural, interconnected systems, providing a sense of shared purpose and collective effort in addressing environmental challenges.

How It Works

1. Setting Up: Ask participants to form a circle and sit on the ground. Ideally, this should be done in a natural space where participants are in direct contact with the Earth. If this is not possible, no problem! any artificial flooring is also good. If participants have mobility challenges, you can use a circle of chairs or any other variation.
2. Ground Rules: Emphasize this as a judgment-free zone with phones silenced, and ask everyone to refrain from taking photos during the experience.
3. The Meditation: Invite participants to close their eyes and, if comfortable, hold hands. Guide them through the meditation, reading slowly and gently through the script, or playing the recording provided.
4. Closing: End the session with thanks to everyone who took part. Consider allowing participants to share some words on how they felt.

Connecting on this essential level makes participants feel whole, grounded, and more connected to the natural world.

Fig. 6.5 Friendly chat after mindfulness meditation in Huilo Huilo Natural Reserve, Los Rios, Chile (2024)

Example Use

The Huilo-Huilo Biological Reserve, located in the Chilean Patagonia, is a globally recognized conservation initiative that combines ecological preservation with community engagement. Spanning over 100,000 hectares of temperate rainforest, this UNESCO-designated site is home to extraordinary biodiversity, including endangered species like the Darwin's frog and the Huemul deer. The reserve not only protects its rich natural heritage but also actively involves local communities through sustainable tourism and environmental education, creating a model for conservation that blends ecological stewardship with social and economic resilience (Fig. 6.5).

At Huilo-Huilo, I facilitated a Design for Conservation (D4C) workshop that focused on innovation for regenerative and responsible tourism as a pathway to conservation. Using D4C's adaptable and participatory tools, we focused on fostering deep connections to place, celebrating cultural heritage, and developing sustainable tourism initiatives that support both conservation goals and local livelihoods. This collaborative approach underscored the potential for tourism to be a catalyst for positive environmental and social change, while curating a meaningful tourism experience.

Reconnection Rituals

Difficulty: Easy
Duration: 30 min.
Materials

- Template printed in A4 or A3 size - one per participant
- Drawing materials (pens, pencils, markers, colors)

Template link

Reconnection Rituals is a simple, engaging tool to help us identify daily activities that reconnect us with our core values, communities, and the natural world. It's all about uncovering the personal or shared moments that make us feel centered and connected to everything around us: people, nature, and ourselves.

Rituals add meaning to life by grounding us in intentional practices, inviting us to pause and reflect on what we value most. They can create a sense of sacredness in everyday actions, transforming even the simplest acts into expressions of appreciation and connection. For example, a morning routine that includes a moment of gratitude, or a walk in nature where you tune into your surroundings, can deepen your sense of connection to life and the world around you. When rituals are practiced in nature, they can draw our attention to the cycles and patterns of the natural world, creating a feeling of harmony and reminding us that we're part of something greater.

How It Works

1. The "Me" Circle: Start by listing actions or moments where you feel true to yourself—these could be times when you think you're living out your values or simply feel at ease. Write these inside the "Me" circle.
2. The "My Communities" Circle: Next, think of times when you feel connected to others without losing your uniqueness. This could be at a family gathering, a community event, or even a casual hangout with friends.
3. The "My Places" Circle: Finally, list the places or experiences that connect you with nature or the world around you. This could include a walk in the park, gardening, or simply watching the stars.
4. Finding Connections: Look at these three circles and see if any elements from the 3 circles could be linked together, maybe an activity or a set of values that are expressed in the 3 levels. Draw lines to connect them between the circles.
5. Story Threads: Pick three of these connections and give them unique names. They can represent themes that emerge from the rituals you've listed. Write a short story about each—these stories can reveal deep insights about your connections.

Fig. 6.6 The reconnection Rituals Canvas. (Gabriela Baron 2021)

6. Explore Insights: Reflect on what you've discovered what have you learned from these rituals?
7. Sharing and Visual Storytelling: You can share these stories with a group or visually express them using the Reconnection Stories Canvas (next tool).

By mapping out and naming your reconnection rituals, this exercise helps you understand what grounds you and brings you closer to others and nature, revealing a deeper sense of unity and belonging (Fig. 6.6).

Example Use

The attached template is an example of how the Reconnection Rituals tool can be used to map meaningful activities that link personal values, social connections, and nature (Fig. 6.7).

In the "Me" circle, personal rituals are listed—activities like tidying up, drinking tea, and admiring beautiful things. These are moments of individual grounding and reflection. Moving outward, the "My Communities" circle includes shared experiences like cooking with loved ones, attending concerts with friends, and exploring art galleries. These reflect social connections and relationships. The outermost

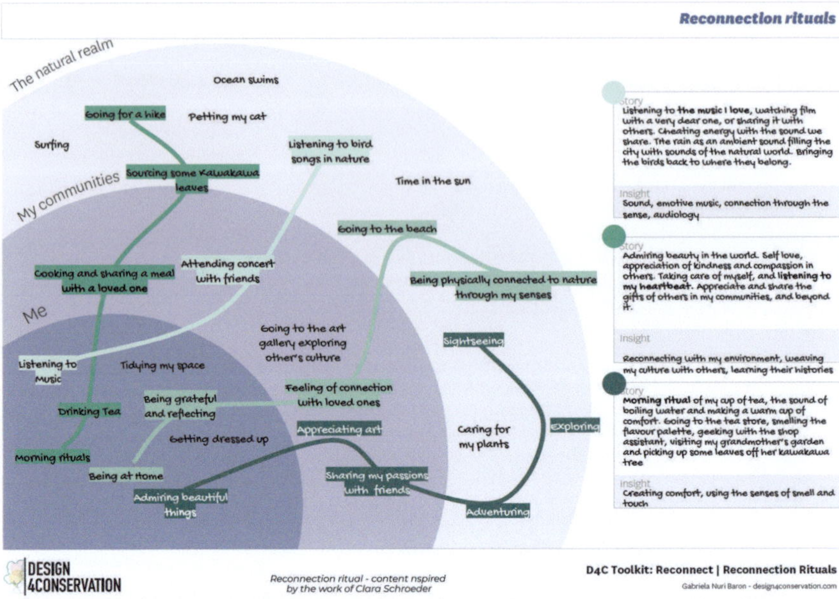

Fig. 6.7 Reconnection Rituals Canvas example. (By Clara Schröder 2021)

circle, "The Natural Realm," captures interactions with the environment, such as hiking, ocean swims, and listening to bird songs, highlighting a deeper connection to the natural world.

What makes this template relevant is the use of lines that link activities across the circles. For example, "caring for plants" connects the personal act of nurturing with the broader connection to nature, while "cooking and sharing a meal" bridges personal and community values. These connections visually demonstrate how actions can flow between the self, others, and the environment.

On the right-hand side, the template includes space for storytelling and reflection. Stories add context, such as how listening to music creates emotional connection or how a morning tea ritual ties into sensory appreciation. Insights derived from these stories will deepen the understanding of what these rituals mean and why they matter.

This example showcases how the tool turns everyday activities into a map of meaningful rituals, providing clarity on how they interweave across different aspects of life. It's a thoughtful way to reflect on what keeps us grounded and connected through the 3 circles.

The next tool will show how we turn these insights into a story.

Reconnection Stories

Difficulty: Medium
Duration: 3 h.
Materials

- Computer or Smartphone
- Template printed in A4 or A3 size - one per participant

Template link

The Reconnection Stories tool helps you explore your positionality and create and share a personal story about your connection to yourself, others, and the world around you. This tool builds on the themes identified in the previous Reconnection Rituals template, encouraging you to transform these themes into a short video, like those that could be shared on social media.

Positionality is the concept of understanding and reflecting on one's own social identity and perspectives as they relate to the topic or context at hand. It recognizes that an individual's background—including factors like race, gender, socioeconomic status, education, and cultural experiences—shapes how they view the world and interact with it. By understanding positionality, participants can approach situations more mindfully, respecting the diversity of experiences and perspectives in collaborative settings. It's a helpful tool in co-design and participatory work, where being transparent about positionality encourages empathy, honesty, and inclusivity.

Sharing a reconnection story within a working group invites others into a more genuine relational space. It counters the often polished, unrealistic images on social media with an authentic, honest narrative. It's a chance to explore your own experiences and share what has shaped you into the person you are today.

How It Works

1. Pick a Story to Share: From the three "threads" identified in the Reconnection Rituals canvas, choose one that stands out to you. Write a script that is around 150 words long and bring this story to life in about a minute. Writing the script is the most critical step in crafting a good story.
2. Plan Your Video: Using the storyboard template, outline the six main scenes that will best tell your story. Keep it simple and heartfelt.
3. Collect Images and Footage: Use photos and videos you already have on your phone, that reflect your journey or capture new moments if you need to(but make sure they are still candid shots and not staged ones).

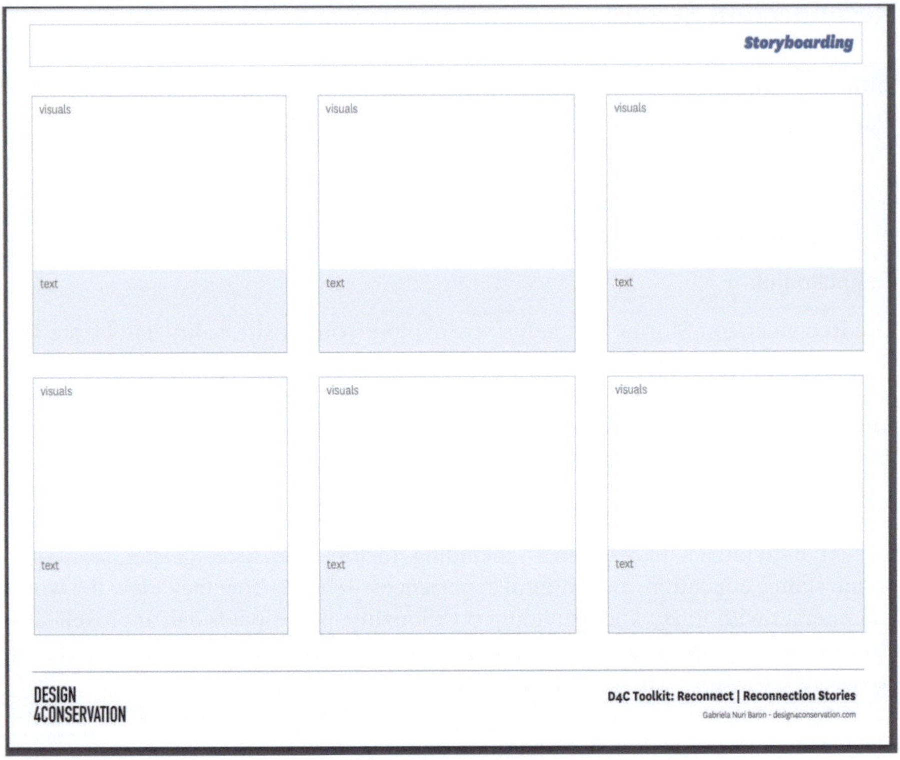

Fig. 6.8 The reconnection stories Canvas. (Gabriela Baron 2021)

4. Edit the Video: Using free (or any) editing software, compile your story. DaVinci Resolve[3] is an awesome free alternative if you have a computer. You can add voiceover, music, and text to emphasize key points and make it more engaging.
5. Share Your Story: Once completed, share your video with your team or on social media if you feel comfortable.
6. The goal here is to create a real, relatable story that highlights your values, your personal growth and your reconnection journey. When working in multidisciplinary teams, this is a meaningful way to show up authentically, explore your positionality, and inspire others to do the same. Watch this inspiring example video by Clara Schröder (2021) (Fig. 6.8).

[3] DaVinci Resolve is a professional video editing and color grading software by Blackmagic Design, widely used in film and TV for its advanced post-production tools and the quality of it's free version.

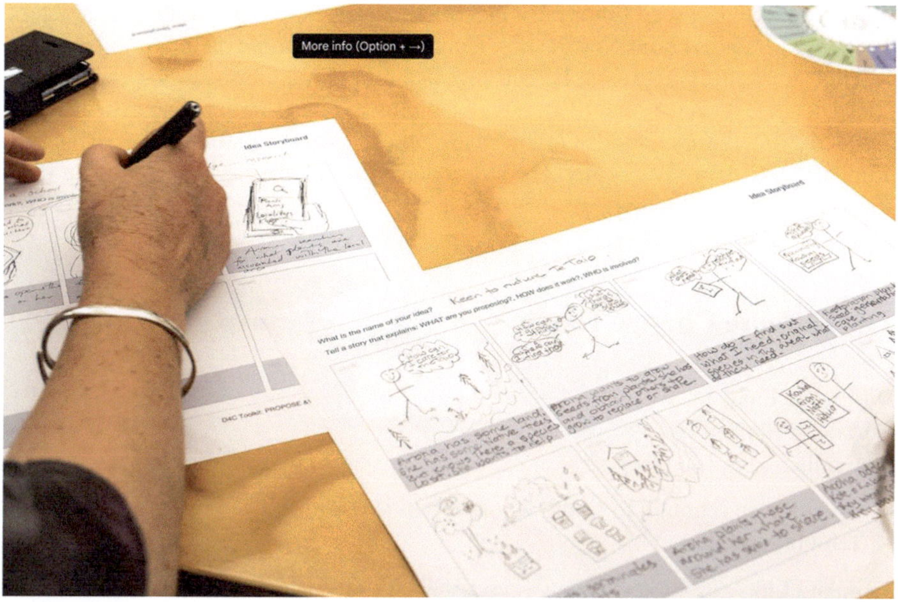

Fig. 6.9 A participant filling out the reconnection stories Canvas. (Photo by Eric Jang 2022)

Example Use

As mentioned widely throughout this book, it is possible to adapt these Canvases for your own project goals. As an example, this storyboard canvas was used instead to communicate hypothetical uses of a new technology, allowing workshop participants (representing diverse partners and stakeholders) to explore how they would interact with it (Fig. 6.9).

This image captures a moment from a two-day workshop held in partnership with Uru Whakaaro[4] during the Auckland Climate Festival[5] in 2022. The focus of the workshop was tackling the challenge: *"How might we create a national seed library to protect native plant diversity, and share knowledge while ensuring Maaori data sovereignty and genuine partnership?"*

In this particular example, participants created storyboards to visually express how a persona group—representing their key stakeholders—might interact with the

---

[4] Uru Whakaaro is a New Zealand-based organization that collaborates with Maaori iwi (tribes), communities, and governmental bodies to promote ecological restoration and cultural well-being. Their services include native forest restoration, ecoscape design, native plant nursery support, provision projects for cultural harvest, zero-waste initiatives, and educational workshops (uru.nz).

[5] The Auckland Climate Festival is an annual month-long event in Tāmaki Makaurau (Auckland) that brings together communities, businesses, and government entities to celebrate and accelerate climate action through various events and initiatives.

concept of a digital seed library. By placing themselves into the story as characters, they told narratives about how the new technology could serve their particular needs.

This storytelling approach helped reveal the practical implications and interactions users might have with the tool. It also allowed us to understand how the proposed digital library could support their goals, foster collaboration, and honor principles like data sovereignty. The storyboards provided a clear and relatable visualization of their perspectives.

The Climate Change Game

Difficulty: Easy
Duration: 40 min.
Materials

- Open Space (enough to fit the whole group in a circle)
- Template printed in A4

Template link

The Climate Change Game is a fun, interactive icebreaker designed to get participants moving, laughing, and reflecting on the interconnected nature of climate change. This activity blends physical engagement with creativity and teamwork, encouraging participants to bond while exploring how environmental challenges ripple through systems and affect communities.

How To Play

1. Set Up: Form a circle with enough room for everyone to stand and move comfortably.
2. Learn the Basics: The facilitator explains the four basic actions:
   - Heatwave: A wave sent in a chosen direction to the next player.
   - Flood: One arm up, bouncing the wave back in the opposite direction.
   - Drought: Both hands up, skipping the next player.
   - Wildfire: Pointing with both hands at another player, sending the wave to them anywhere in the circle.
3. Practice Rounds: Run through a few practice rounds to ensure everyone is comfortable with the actions.
4. Create New Actions: Split into small groups and brainstorm additional climate-related actions (Maximum 5 groups). Examples include:
   - Cyclone: Players spin, and the wave moves outside the circle.

- Panic: Players mix positions in chaos.
- Blizzard: A targeted player freezes until another unfreezes them.
- Denial: All participants turn around with their backs towards the circle, moving the game to the outside of the circle—which is a lot harder to follow—until someone says "science!" in their turn, so that all participants face each other again and continue playing.

5. Play: Have each groups explain and demonstrate their new actions and incorporate them into the next rounds. Players use any action on their turn, and mistakes result in elimination.
6. Finish: The last remaining player wins the game.

After playing, regroup to reflect on the experience. Discuss how actions can also "mitigate" or "adapt" to climate change and brainstorm positive actions for inclusion in future games.

End with unifying actions like "Community" (passing high-fives around the circle) and "Hope" (a giant group hug). These moments emphasize our shared responsibility and highlight the importance of collaboration in addressing global challenges.

This game isn't just about fun—it's about creating connections, fostering creativity, and reminding participants of the collective effort needed to protect our planet (Fig. 6.10).

Example

This game is AMAZING and always brings people together. Seriously, don't hesitate to try it with adults, it's just as fun for them as it is for kids (sometimes even more so!). I recommend using it on a day when you want to break the ice or energize a group. Depending on the group, it might feel a bit awkward at first, but once everyone gets into it, you'll feel the energy shift. People start laughing, opening up, and showing their true selves. And when participants create their own actions, it's even better—you're encouraging agency and co-creation right there in the game!

Before or after the game, it's really important to frame its purpose. Climate change is a serious and pressing challenge we're all facing. It's no longer a distant issue, and it's impacting our own communities in ways we can't ignore. The goal of this game isn't to dismiss or make light of this reality, but to create a space where we can talk about it in a more positive, collaborative way. It helps us reframe the problem so we can focus on how we can work together to adapt and make a difference (Fig. 6.11).

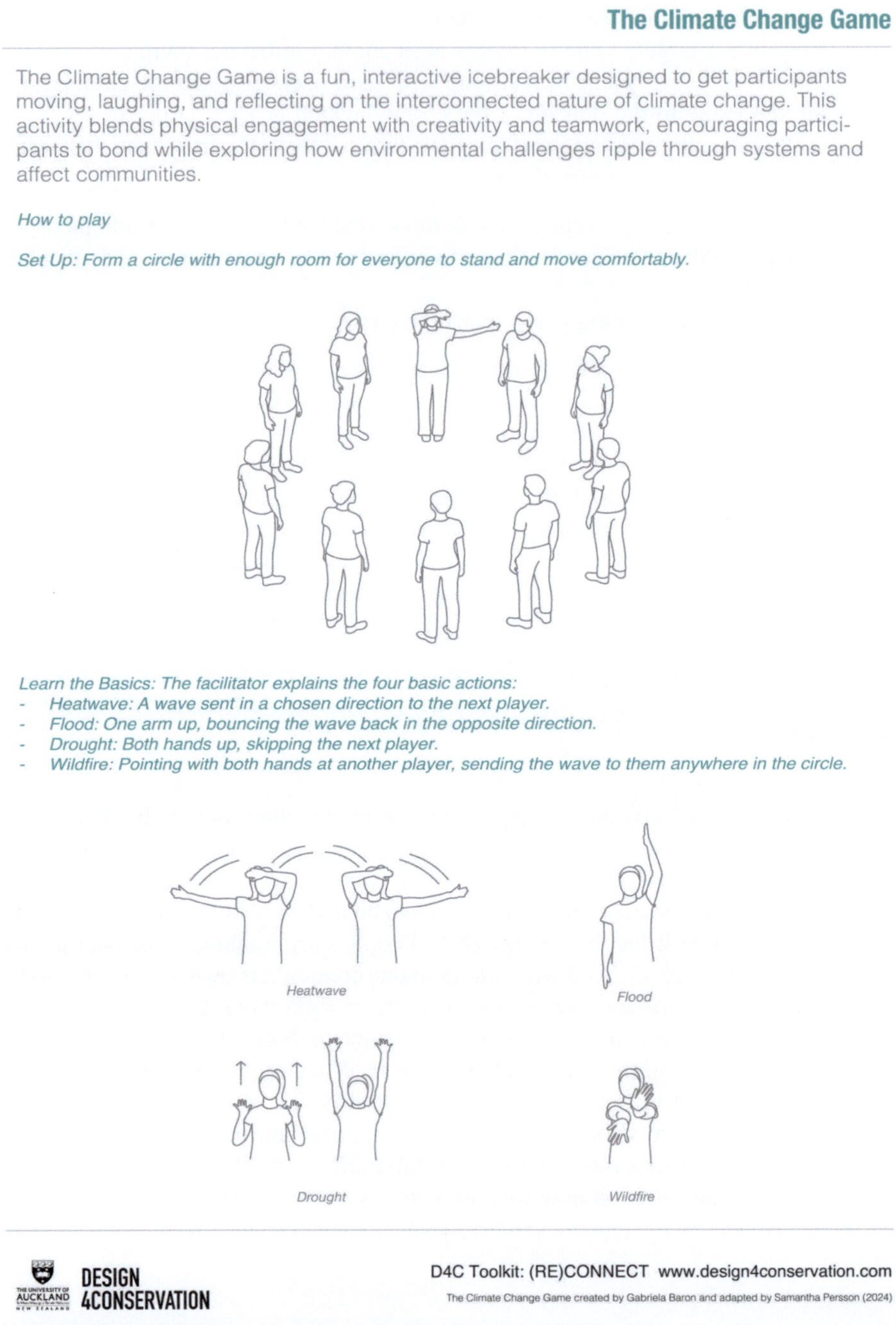

Fig. 6.10 The climate change game template that can be used as a guide for facilitation. (Gabriela Baron and Samantha Persson 2024)

Fig. 6.11 Reflecting on the climate change game with a group hug at Earthbeat Festival (Earth Beat Festival is a five-day event in Ātiu Creek Regional Park, Auckland, celebrating music, arts, and sustainability with workshops and family-friendly activities. (earthbeatfestival.com)) 2023 in Aotearoa, New Zealand

Chapter 7
Understand

Gabriela Nuri Baron

The Understand phase is all about diving deep into environmental challenges by setting aside preconceived ideas and approaching issues with an open, systemic view. Environmental problems are quite complex, so they demand a mindset that can embrace ambiguity and look at the larger web of connections: how people, land, species, and ecosystems influence one another.

In this phase, it's crucial to listen to a wide range of perspectives and actively recognize that valuable expertise often lies with those who have a close, lived relationship with the environment. These might be local communities, Indigenous groups, or those who work directly with the land and species impacted. Their knowledge provides insights that are essential to forming an unbiased understanding of the problem.

We can also think of this phase as a time to *unlearn* assumptions that may limit our openness to solutions. This means setting aside any fixed ideas we may have about what the solutions should look like and instead letting the process guide us.

Let's explore the tools of this phase.

G. N. Baron (✉)
Faculty of Engineering and Design, University of Auckland – Waipapa Taumata Rau, Auckland, New Zealand
e-mail: g.baron@auckland.ac.nz

G. N. Baron (ed.), *Design for Environmental Conservation*,
https://doi.org/10.1007/978-981-96-9574-4_7

Environmental Conservation Challenge

Difficulty: Medium
Duration: 1 h.
Materials

- Pens, markers
- Template printed in A2 or A3 size—one per group

Template link
Video explanation link

This canvas will be the backbone of any Design for Conservation (D4C) workshop, helping participants establish clear and shared agreements on the challenges they're tackling together. The aim here is to make sure everyone is on the same page and looking at the *right* problem before diving into solutions.

This tool walks the group through a structured process to help diagnose the problem, make sense of the information they already have, and identify what's missing. It's also designed to guide teams in defining clear, actionable design challenges that emerge from this collective understanding. And remember, the goal isn't perfection on the first try; it's about creating a foundation that can evolve as the project unfolds and insights deepen.

How It Works

1. Project Goals:
   - Grand Challenge: Start by agreeing on the big, global, conservation goal, like *"restoring coastal ecosystems."*
   - Specific Problem: Next, refine it to reflect the local impact of this grand challenge. For example, *"increase local plant biodiversity"*.
   - Project Goals: Specify concrete goals, like *"carry out riparian planting"*, *"provide native seeds to the community"* or *"raise local awareness on coastal conservation."*
   - Link to UN Sustainable Development Goals: Mapping goals to relevant SDGs helps align your project with global conservation targets (this is absolutely optional and will not affect the definition of the problem).
2. Context:
   - Cultural Obligations: Respect local traditions, values, or Indigenous knowledge. How will the project recognize and honor this?
   - Connection to Place: Think about physical and environmental factors affecting the area, like "these coastal communities are particularly vulnerable to rising sea levels."

- Voices and Perspectives: Capture a range of views, including differing or conflicting ones, to get a well-rounded perspective.
- Requirements: Note specific constraints, such as budget limitations or environmental regulations that may affect the project.

3. Main Stakeholders:
   - Human, Non-human (or more-than-human), and Institutions: Identify who is most affected by this problem and who should have a voice in solving it. This could include local residents, wildlife species, NGOs, or government agencies.
4. Design Challenges:
   - Frame clear, actionable questions that make it easier to approach the problem from different angles using a "*How Might We*" structure. For instance, *"How might we encourage native planting in community spaces?"*
5. Impact:
   - Imagine the outcomes if your project is successful. Define what success means in terms of both qualitative and quantitative indicators, like *"an increase in native bird sightings"* (quantitative) or *"better mental health from hearing birdsong more often"* (qualitative).
6. A Living Document:
   - Treat this document as an evolving tool. Revisit it to adjust goals, refine challenges, or incorporate new insights as your project progresses.

Using this canvas allows everyone involved to build a collaborative roadmap for understanding and addressing the conservation challenge at hand. It's all about setting a solid foundation and being open to ongoing learning and adaptation. Check out the video explanation here, where we provide examples of make the best of this tool (Fig. 7.1).

Example Use

For this project, we pinned the Conservation Challenge onto a board, treating it as a living document. While initial agreements on problem definition provided a foundation, we continuously added to and refined it as the work evolved. This flexibility ensured the canvas remained relevant and responsive to emerging insights (Fig. 7.2).

As you engage with this Canvas, consider adapting it to fit your project's unique needs. You might refine sections, remove those that feel unnecessary, or introduce new ones that better align with your goals and constraints. For instance, in the image below, you'll see how I customized this Canvas for a sustainable mobility workshop—demonstrating how it can be tailored to different contexts (Figs. 7.3 and 7.4).

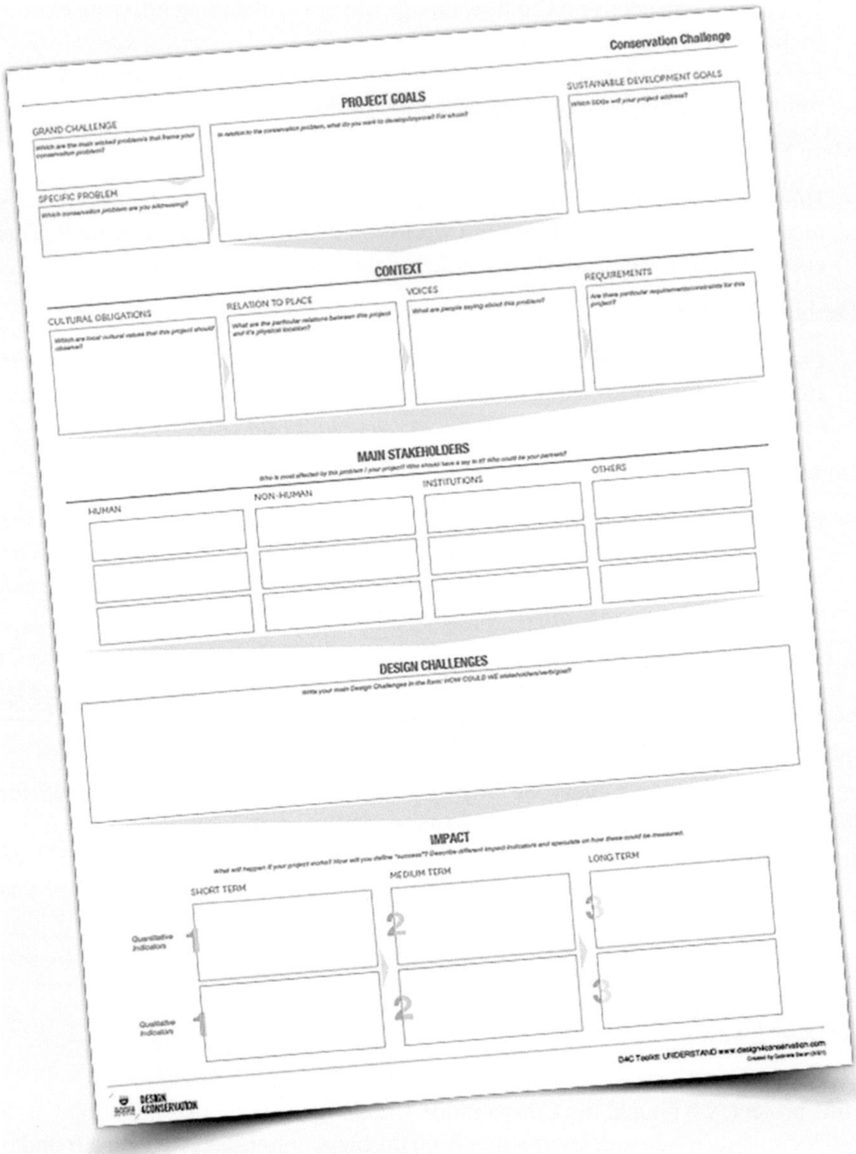

Fig. 7.1 The conservation challenge canvas. (Gabriela Baron 2021)

Stakeholder Characterization

Difficulty: Medium
Duration: 1–3 h.

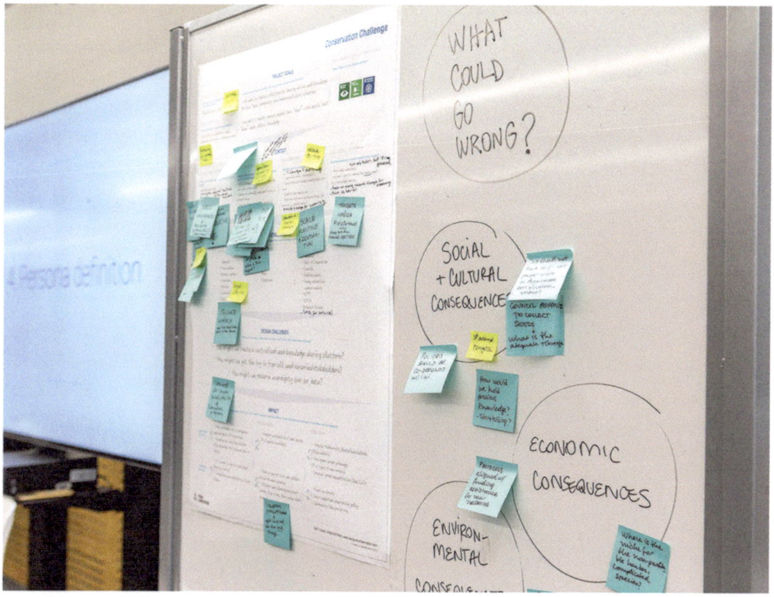

Fig. 7.2 Conservation challenge canvas in use. (Gabriela Baron (2022). Photo by Eric Jang (2022))

Materials

- Pens, markers
- Template printed in A4 or A3 size—as many as needed to be shared in groups.

Template link

The Stakeholder Characterization Canvases are a practical way to help teams understand their stakeholders more deeply, building on the insights gathered in the Conservation Challenge Canvas. This tool uses ***"personas"***—fictional characters representing real stakeholder types—to capture the main traits, goals, and needs of the people, non-human species, and organizations connected to the project.

Types of Personas

1. Human Personas: These are fictional profiles representing individuals within your stakeholder groups. Each human persona should include relevant details like age, role, location, motivations, and attitudes toward conservation. This helps teams understand what drives different people in the community and how they might interact with the project.
2. Non-Human Personas: These represent species, ecosystems, or other natural elements affected by the project. Non-human personas include scientific or common names, habitats, needs, and specific vulnerabilities, helping teams integrate a multispecies approach. Because these personas will not have a *voice* in the projects, team members and specialists can be their advocates. Just like non-

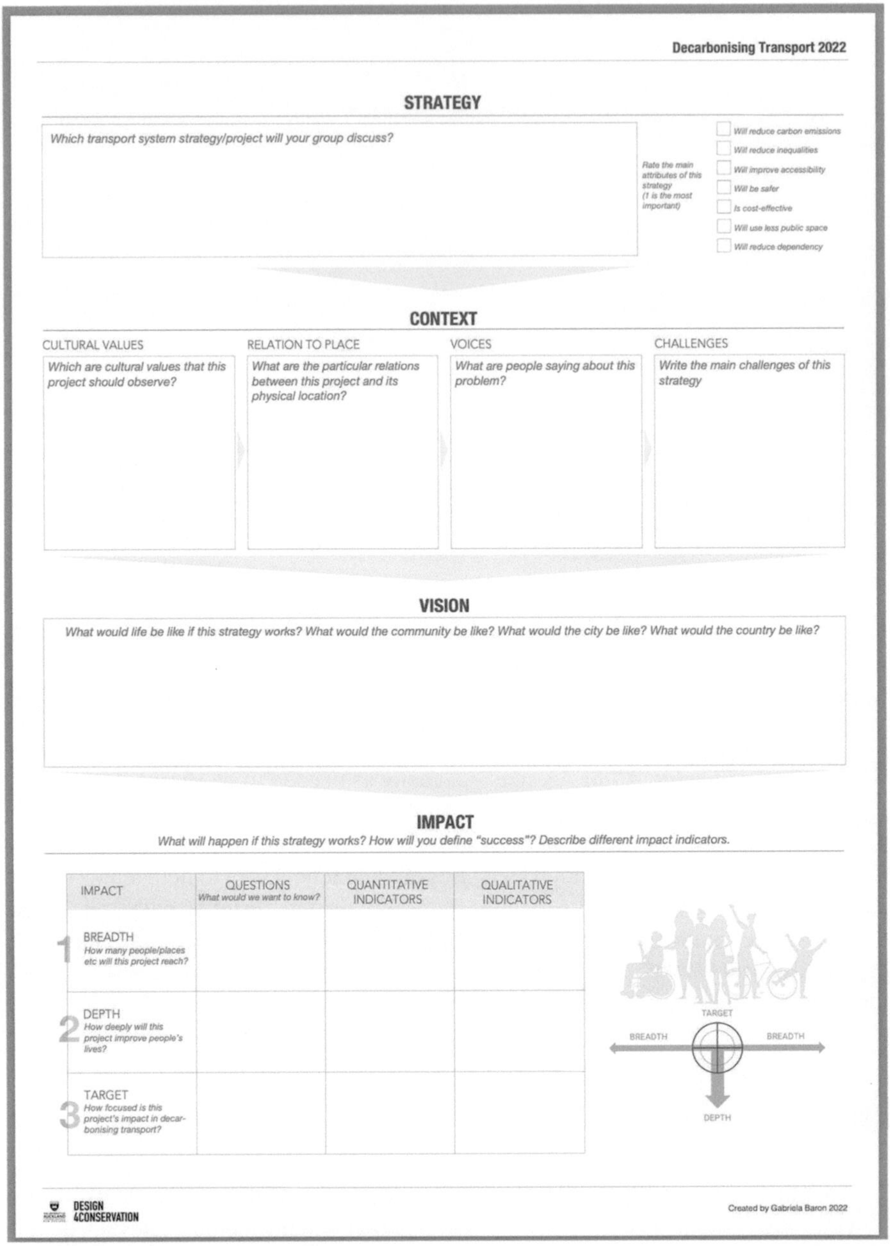

Fig. 7.3 A variation of the conservation challenge canvas created for the "Decarbonising Transport" Workshop. Gabriela Baron (2022)

Fig. 7.4 Snapshot of "Decarbonising Transport" workshop organized by the Transportation Research Centre at the University of Auckland (2022)

humans are gaining legal personhood in some countries, like Taranaki Maunga (Mount Taranaki) in New Zealand,[1] are seen as ancestors by Indigenous peoples, and deserve the same respect and obligations as any human client.

3. Organizations: Organizational personas represent real or fictional organizations that may play a role in the project. These profiles outline the organization's type, role, motivations, resources they may contribute, and level of commitment to the project's goals. This information helps teams identify potential partnerships and leverage resources.

How It Works

1. Define your key stakeholders and decide how many personas to create: Each persona should embody one type of stakeholder (e.g., local residents, native plant species, government agency). You can start with the stakeholders listed in the Conservation Challenge Canvas. Ideally, you'd have at least 6 personas, and up to as many as you need.
2. Fill out the template by answering key questions:
   - For human personas, consider their goals, needs, and relationship with conservation work.

[1] In 2017, Mount Taranaki (Taranaki Maunga) in Aotearoa New Zealand was granted legal personhood, following similar recognition of the Whanganui River. This approach aligns with Indigenous worldviews, where nature is seen as kin rather than a resource.

– For non-human personas, think about their ecological role, needs, and any special vulnerabilities.
– For organizations, outline their motivations, potential contributions, and the role they could play in the project.

Risks of Using Personas

Creating user personas can be incredibly helpful, but it comes with risks. If based on incomplete data or assumptions, personas might misrepresent users, leading to biased results. They can also oversimplify complex behaviors, reinforce stereotypes, or exclude marginalized groups. Teams might rely too heavily on personas, treating them as absolute truths and neglecting direct user engagement. To avoid these pitfalls, ensure personas are grounded in solid research, represent diverse experiences, and are revisited regularly as user needs evolve. Teams can overcome this by using personas as a starting point, not the whole picture, and always pair them with real-world user testing as the project unfolds (Fig. 7.5).

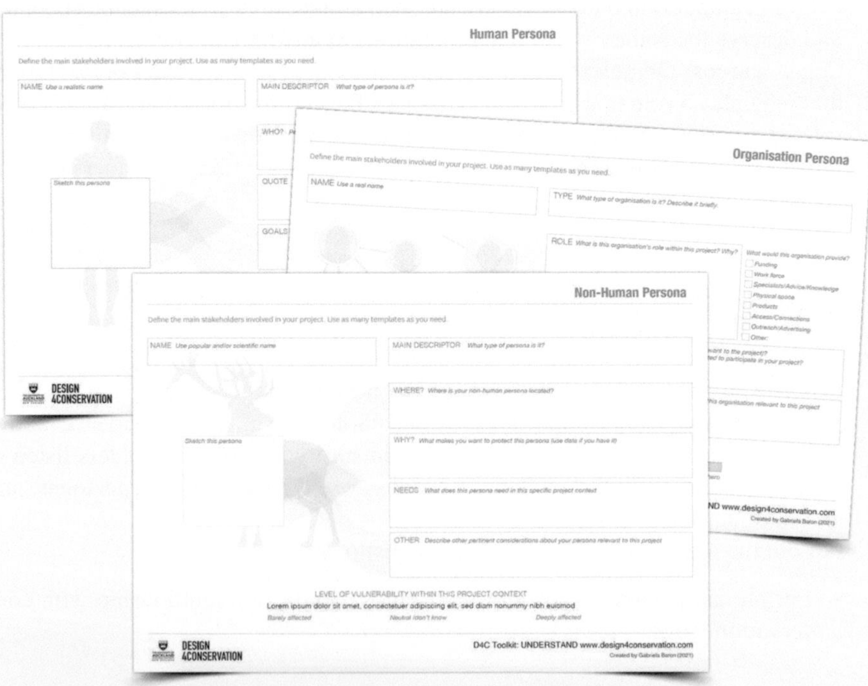

Fig. 7.5 The persona canvases showing 3 types of persona types. (Gabriela Baron 2021)

Fig. 7.6 Participant using the non-human persona canvas to describe the Pūriri Moth, endemic to the North Island of New Zealand. (Photo by Eric Jang 2022)

Example Use

Imagine your project involves restoring a local wetland. You might create a human persona for a nearby resident who values the area for recreation, a non-human persona for a native bird species affected by habitat loss, and an organizational persona for a conservation NGO interested in partnering on the project.

This tool ultimately ensures that the team can empathize with and advocate for each stakeholder's needs throughout the project, leading to more balanced and inclusive design decisions (Fig. 7.6).

Stories as Insight

Difficulty: Easy
Duration: 1 hrs.
Materials

- Pens, markers
- Template printed in A4 or A3 size—one per story

Template link

The Stories as Insight tool is a way to tap into the deep knowledge embedded in traditional stories. Through storytelling, traditional knowledge isn't just preserved;

it's lived and felt. These stories, shared by community members, elders, or relevant stakeholders, often carry generations of wisdom and understanding about a specific place or environmental challenge. By listening to these stories, you can gain rich detail and insights that may not come up in other research methods like interviews or focus groups.

How It Works

1. Approach Respectfully: Choose someone you know personally, or get introduced by a relevant friend or stakeholder. If this person is from a different cultural background, approach them with cultural sensitivity and respect.
2. Request a Story: Ask if they would be willing to share a story related to your conservation challenge. You might start with questions like:
   - *"Is there a traditional story that was told to you by someone important to you?"*
   - *"What's the message of this story, and why does it stand out to you?"*
   - *"Who do you think should hear this story, and why?"*
3. Reflect: Once you've heard the story, take some time to think about its messages. Sketch an image related to the story if that helps you connect to it more visually.
4. Unpack: With your team, discuss the story's meaning and key insights. Write down what you've learned and consider how these lessons might inform your project.

This process is about capturing and respecting knowledge passed down through storytelling, which adds depth to your understanding and helps root your project in the cultural and environmental context of the community (Fig. **7.7**).

Agile Interviews

Difficulty: Easy
Duration: 1 hrs.
Materials

- Pens
- Template printed in A4 or A3 size—one per interview

Template link

The Agile Interviews template helps teams gain a deeper understanding of the project and its challenges by engaging directly with the people most connected to the land or issue at hand. This could include community members, Indigenous leaders, scientists, academics, or others with expertise or lived experience.

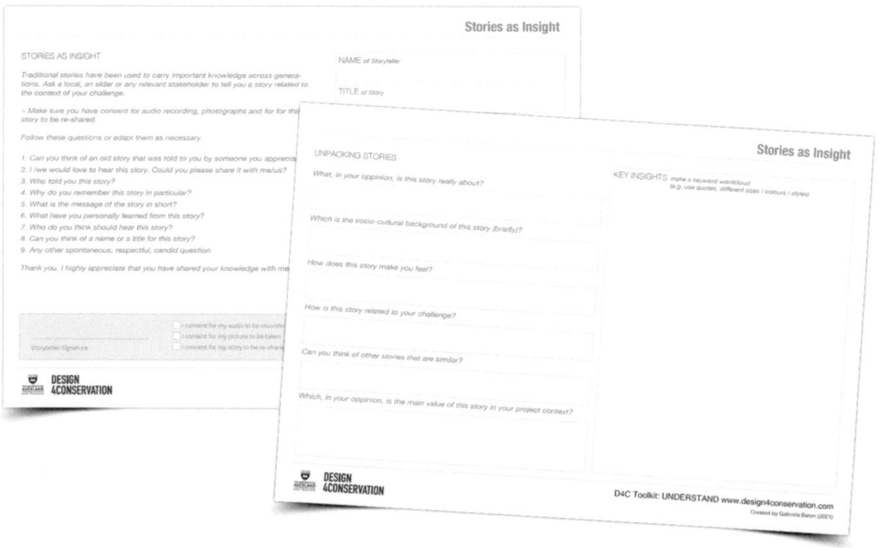

Fig. 7.7 The Stories as insight template. (Gabriela Baron 2021)

How It Works

1. Be Genuine and Respectful: Approach each interview with honesty, sensitivity, and respect. Show genuine interest in the interviewee's perspective, and review ethical guidelines for research if you're unfamiliar with them.
2. Prepare Thoughtful Questions: Plan questions ahead of time, starting with broader topics to help ease into the conversation before moving to specifics about the project. Let the interview flow naturally, sometimes the best insights come from unexpected directions.
3. Prioritize Comfort and Accessibility: Ensure the interviewee is comfortable with the setting and time. Offer refreshments if possible, and consider covering any travel expenses. If interviewing Indigenous participants, consider giving a small gift or *"koha"* (for Maaori in Aotearoa) as a gesture of respect.
4. Be Transparent: Clearly introduce yourself, explaining your background, the purpose of the research, and any organizations involved or funding the project. Transparency builds trust.
5. Confidentiality: With consent, record the interview or take notes, and store this information securely, respecting any agreed confidentiality terms.
6. Show Gratitude and Stay Connected: After the interview, think of ways to give back to the interviewee or their community. Share the results of your research when it's completed, cite contributors properly, and maintain a respectful relationship.

This approach makes interviews not only informative but also a positive and respectful experience for everyone involved, building long-term relationships of trust (Fig. **7.8**).

Fig. 7.8 The Agile interviews template. (Gabriela Baron 2021)

Immersive Field Trip

Difficulty: Medium
Duration: 1–8 h.
Materials

- Pens, markers
- Template printed in A3 size—one per trip

Template link

This tool is designed to help you gain a firsthand, in-depth understanding of the area most affected by the conservation challenge you're working on. By visiting the place, observing, and experiencing it directly, you can connect with it in a way that goes beyond just facts.

How It Works

1. Arrive with Humility and Respect: Approach the place and its people with genuine curiosity and sensitivity. Keep in mind that you're a guest, so be respectful and consider reviewing research ethics before your visit.
2. Document the Facts: Write the name, place, and date to leave a written record of the experience, along with a detailed map of your trajectory, stops, etc.

3. Engage All Your Senses: Spend time exploring on your own and tune in to your surroundings. Notice the sights, sounds, and scents of the place—what emotions or reactions do you experience? Take off your shoes, touch the plants, swim in the water, or feel the earth if appropriate, to deepen your sensory connection with the environment.
4. Create a Mood Board: Capture the essence of the field trip with a visual mood board. This can include photographs, sketches, keywords, and personal notes. The goal is to create a snapshot of the experience that reflects both what the place looks like and how it made you feel.
5. Document Respectfully: With consent, you can record or photograph. Remember that your presence should have a positive impact.
6. Leave the area better than you found it: Whether through formal volunteering or simple acts of care, find a way to give back. This could mean picking up litter, participating in conservation efforts like tree planting, or even helping with tasks such as washing dishes after a meal.

This tool is all about fostering a deeper connection to the place and people most affected by the problem you're tackling, making sure your insights are grounded in humility and respect (Fig. **7.9**).

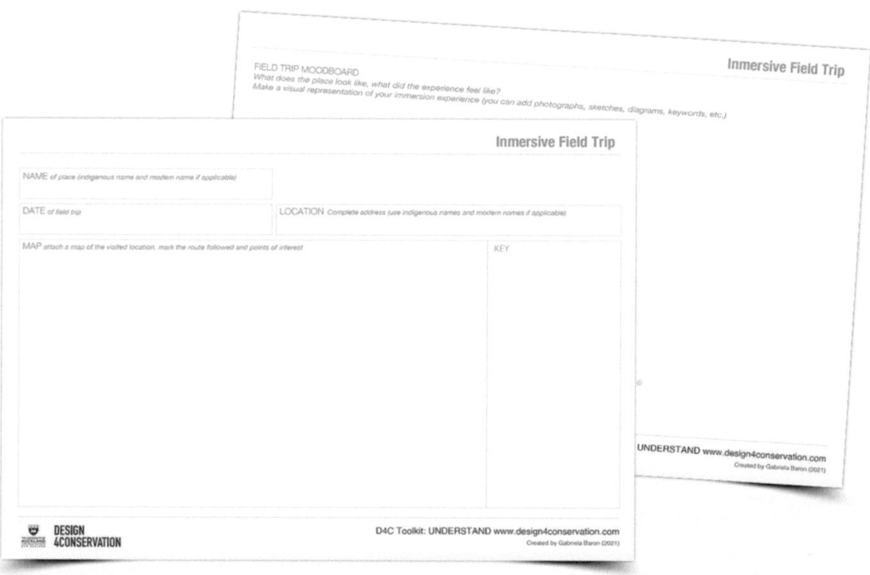

Fig. 7.9 The Immersive field trip template. (Gabriela Baron 2021)

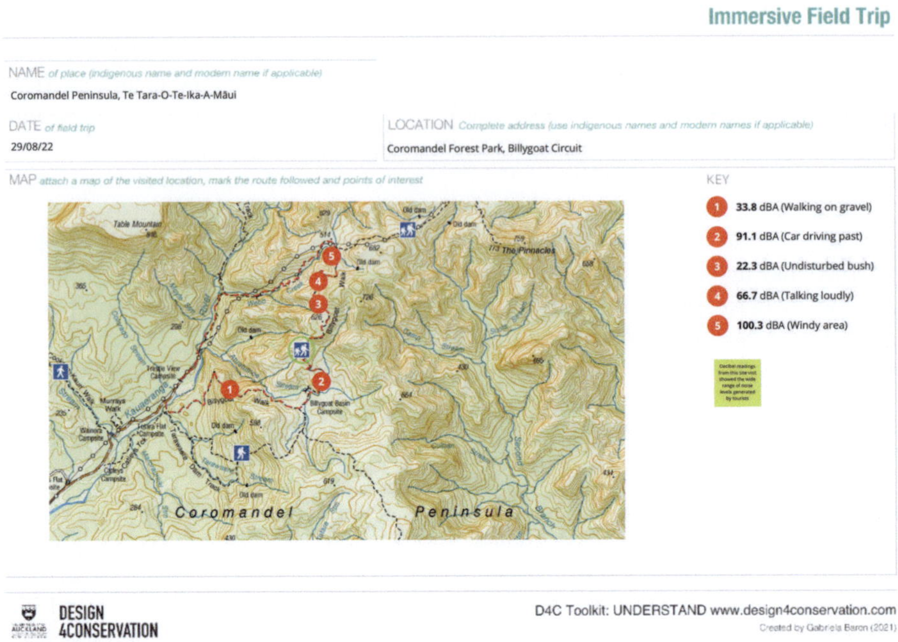

Fig. 7.10 The Immersive field trip canvas used by students for the Atmos project. (Authors: Siena Thompson, Isobel Mullins, Isobel Caldwell, and Wayne Poon (2022))

Example

The Atmos project (2022) addresses the issue of noise pollution caused by tourism and its detrimental effects on wildlife and biodiversity. The proposed solution is a series of interactive light sculptures, designed as Pouwhenua,[2] that visually respond to noise levels, shifting from white to red to raise awareness among tourists about maintaining quiet environments.

For this project, students used the field trip Canvas to document a hiking trip where they measured human noise pollution in decibels (Fig. 7.10).

[2] Pouwhenua are carved wooden posts used by Māori to mark territorial boundaries, commemorate ancestors, or convey important cultural narratives. They are often placed in significant locations and serve as symbols of connection to the land, whakapapa (genealogy), and identity.

Chapter 8
Propose and Validate

Gabriela Nuri Baron

The "Propose and Validate" stage is all about problem-solving. The following tools will help you gain inspiration from natural systems, and gain inspiration from any field, to arrive at sustainable solutions. This is an exciting and creative stage where the actionable problem-solving begins! Teams will take the insights and knowledge gathered in the earlier stages and turn them into tangible solutions that will impact conservation efforts. You will prototype ideas rapidly, validate them with your stakeholders, learn from mistakes, and evolve your proposal.

Let's dive into the tools.

Inspired Ideation Lotus

Difficulty: Medium
Duration: 90 min.
Materials

- Pens, markers, sticky notes
- Template printed in A0 or A1 size - one per group

G. N. Baron (✉)
Faculty of Engineering and Design, University of Auckland – Waipapa Taumata Rau, Auckland, New Zealand
e-mail: g.baron@auckland.ac.nz

G. N. Baron (ed.), *Design for Environmental Conservation*,
https://doi.org/10.1007/978-981-96-9574-4_8

Template link

The Inspired Ideation Lotus is a brainstorming tool designed to help teams break down successful examples from other fields and adapt their best characteristics to create fresh ideas for conservation challenges. It's a hands-on, visual method that uses a nested *lotus* diagram format to organize thoughts and inspire new solutions through structured ideation (Fig. 8.1).

How It Works

1. Set the Challenge: Begin by writing your main conservation challenge, or a specific part of it, in the center of the poster.
2. Discuss Key Strategies: Around the center, you'll find 8 D4C strategies. Discuss how each of these strategies might address the challenge. If your team feels other strategies would be more suitable, you can substitute them, these strategies are just a suggestion to help you understand the template.
3. Add Inspirational Examples: For each strategy, find an inspiring example—usually from a different field—that demonstrates how this strategy can be applied. Use sticky notes to write down these examples on the colored boxes surrounding each strategy. In this step it's key to think outside the box, maybe you are working towards plant biodiversity but your strategy is *"easy to trust/transparent"* and then the inspiration case could be *"Uber rides"* because the app gives you all the info you need to know in a transparent way. Plants and Uber don't seem to have anything in common, but trust the process!
4. Identify Key Characteristics: Analyze what makes each example effective and note the top 8 qualities that stand out. Write these characteristics around each example. Don't worry, you do not need to fill out all the boxes, but try to go for quantity.
5. Combine and Create: Mix and match the best qualities from the examples you discussed, using them as building blocks to develop 3 new solution ideas for your conservation project. Briefly describe each idea in the spaces provided on the poster.

This process breaks down the requirements of complex challenges into manageable parts, enabling the team to explore a variety of angles and merge the best characteristics of inspirational examples into creative, actionable ideas for their project (Fig. 8.2).

Example

Sea Gardens is a marine regeneration project on Auckland's waterfront, restoring biodiversity in the Hauraki Gulf through artificial reef structures that support mussel reefs. It reconnects people with the harbor via interactive sculptures, educational displays, and partnerships with Maaori artists, aiming to enhance water quality, revive ecosystems, and promote sustainable seafood harvesting (Fig. 8.3).

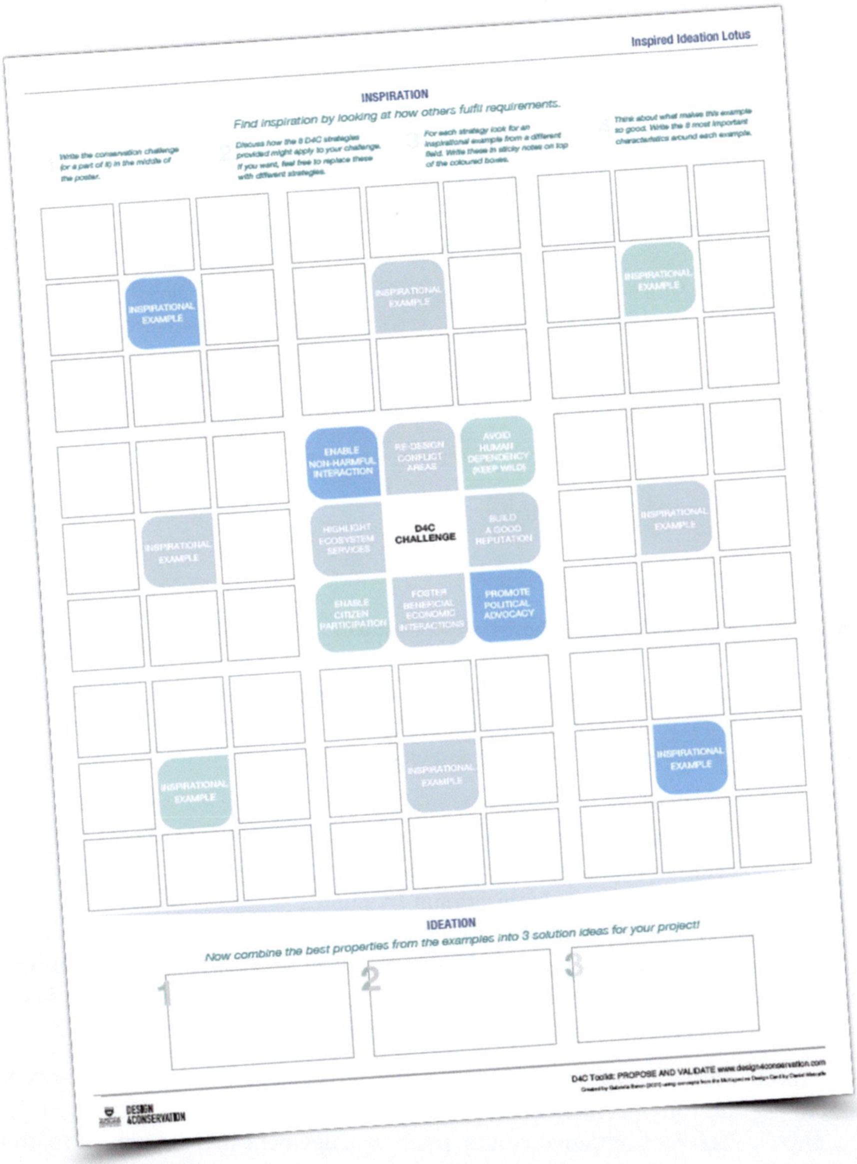

Fig. 8.1 The inspired ideation lotus canvas. (Gabriela Baron 2021)

Fig. 8.2 The inspired ideation lotus canvas in use. (Photo by Eric Jang 2022)

The Sea Gardens project applied the Inspired Ideation Lotus method to refine its approach to marine regeneration by integrating diverse strategies and best practices from different fields. Here's how the framework was used:

1. Setting the Challenge

   The central conservation challenge focused on restoring biodiversity in the Waitemataa Harbor by engaging local communities through interactive elements.
2. Applying D4C Strategies & Inspirational Examples

   The team discussed how applicable[1] the 8 D4C strategies that surround the challenge were for their project. In the example, the Sea Gardens project decided to explore the default Design for Conservation (D4C) strategies, drawing inspiration from successful examples in different fields. For example:

   - **Enable Non-Harmful Interaction:** The project could foster human-ocean connection through interactive sculptures, similar to how the *Experiencing Marine Reserves* initiative works, public engagement installations in marine reserves encourage learning and connection.
   - **Highlight Ecosystem Services:** By collaborating with Māori artists and local stakeholders, Sea Gardens could ensure its ecological restoration efforts align with both nature-based solutions and cultural values, much like *Te Wānanga*

[1] Note that if these strategies were unsuitable, you could create your own strategies. For example: *"Easily scalable", "Promote transparency and trust", "Data-driven"* or *"Place-based"*.

Fig. 8.3 Ideation lotus leading to the sea gardens project (By Cara-Sophia Andres, Elouise Kidd, Benjamin Jarvis, and Jade Nolan, 2023)

- **Re-Design Conflict Areas:** Artificial reefs might mimic natural mussel beds, taking inspiration from living walls projects that have improved water quality and marine biodiversity.
- **Avoid Human Dependency:** The reef structures might support self-sustaining ecosystems, inspired by low-maintenance ecological restoration projects that regenerate over time with minimal intervention.
- **Build a Good Reputation:** Public engagement and education could play a key role in promoting the project's impact, following the example of youth-focused climate campaigns that use storytelling and media to inspire change.

3. Identifying Key Characteristics

   On the 8 spaces around each inspirational example, key success factors were identified, including:

   - Ecosystem scale action (big-picture impact)
   - Cultural inclusion (working with Māori artists and kaitiaki)
   - Public connection (education and awareness-building)
   - Simple, focused design (reef structures inspired by nature)
   - Long-term ecosystem impact (biodiversity restoration over generations)

4. Generating New Solution Ideas

   Once the team completed as many of the boxes as possible, they had a large range of inspirational ideas. Then, they discussed how to converge into 3 solution ideas by combining the best characteristics of each inspirational example:

   1. **Under the Sea:** A live-streaming underwater camera to provide real-time insights into the harbor biodiversity and water quality improvements.
   2. **EcoHull:** A boat hull protection device that minimizes damage to marine ecosystems and supports natural regeneration.
   3. **Seagarden:** An ecological health indicator sculpture that educates the public on biodiversity and water quality.

The Inspired Ideation Lotus helped the Sea Gardens project synthesize best practices from marine conservation, Indigenous knowledge, and interactive design to create a compelling, community-driven approach to ecological restoration.

As you can see, this is much more powerful than traditional brainstorming. By leveraging cross-pollination—borrowing ideas from successful solutions in other fields—the team was able to develop more creative, effective, and scalable solutions.

Now, the next tool can help teams choose between these 3 ideas in an equitable way.

Idea Selection Checklist

Difficulty: Medium
Duration: 30 min.
Materials

- Pens
- Template printed in A4 size—one per person

Template link

The Idea Selection Checklist is a tool to help teams evaluate and choose the strongest ideas from a brainstormed list. After generating ideas with the Inspired Ideation Lotus Canvas, this checklist provides a way to organize and assess them based on key criteria, ensuring that every idea gets a fair review.

How It Works

1. Choose Best Ideas: Start by picking your top 3 ideas and give each one a short, clear name.
2. Agree on Selection Criteria: For a fair, inclusive process, agree on the most important criteria as a team. The template provides a series of criteria that may fit any project, but feel free to adjust the checklist to better fit your project's unique goals and context. Defining new criteria should be done in a collective way, including all voices, reflecting the most important requirements for your project.
3. Apply the Selection Criteria: Individually, use the checklist provided to discuss how well each idea meets the following attributes:
   - Adequate/Sustainable: Does the idea achieve the project's goals? Is it scalable, sustainable, and tailored to the local context?
   - Creative/Innovative: Is it clear, adaptable, and likely to capture people's interest? Does it inspire others or challenge conventional thinking?
   - Magical: This is the final point and should only be awarded if the idea has a *wow* factor that resonates deeply or inspires change.
4. Scoring: Add up points for each attribute your idea meets. Add up the scores from all participants. In the end, the idea with the highest score is the one you'll develop further as a team.

This tool is all about making sure the chosen idea is desirable, feasible, and inspiring while allowing the team to collaboratively set the criteria that matter most (Fig. 8.4).

Example Use

The Paper Bird Project addresses speciesism by fostering empathy and understanding toward the Tarāpunga (Red-Billed Gull), a species often overlooked in Aotearoa's conservation efforts. Through literature designed for children in

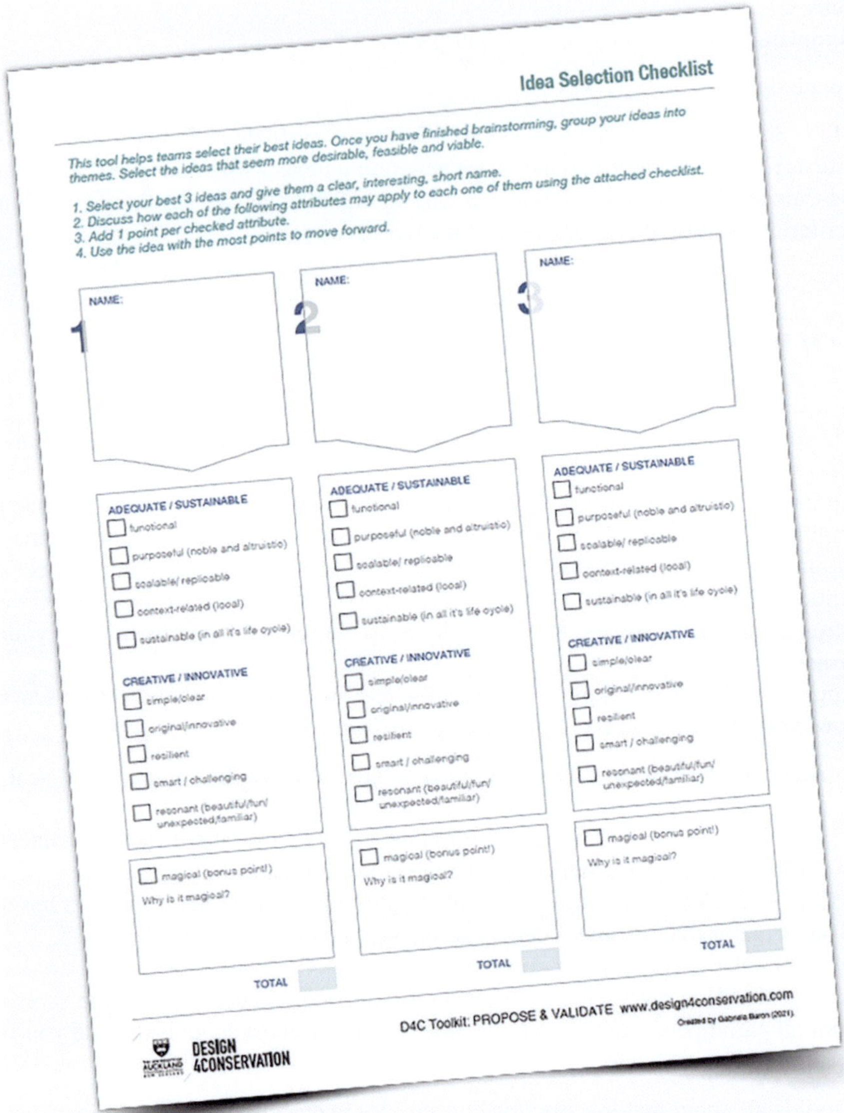

Fig. 8.4 The idea selection checklist tool. (Gabriela Baron, 2021)

Tāmaki Makaurau (Auckland), the project seeks to reshape perceptions and nurture compassion from an early age. By promoting meaningful connections with wildlife, it aims to break the cycle of speciesism for future generations (Figs. 8.5 and 8.6).

Fig. 8.5 Idea selection checklist example for The Paper Bird Project by Stephanie Townend

Fig. 8.6 High fidelity prototype for The Paper Bird Project by Stephanie Townend (2023)

Mitigating Bias

By collectively establishing selection criteria and using an equitable voting system, all ideas are given a chance to be considered, minimizing bias and ensuring diverse perspectives are valued. The Idea Selection Checklist enhances this process by offering a structured framework to evaluate ideas based on agreed-upon criteria like sustainability, creativity, and impact. This approach not only promotes fairness but also helps teams identify the most inspiring, feasible, and transformative ideas to move forward together.

There are many other ways to do this. The example below shows a more fluid approach to idea selection, by using dots to vote, participants can ponder ideas following collectively established criteria (Fig. 8.7).

This image shows the results of a collaborative activity aimed at prioritizing strategies for decarbonizing transport. The diagram is structured as a two-axis matrix, with "Short Term" to "Long Term" on the horizontal axis and "High Carbon" to "Low Carbon" on the vertical axis.

Workshop participants used sticky notes to propose various strategies for reducing carbon emissions in transportation, such as *"Shared Cars," "New Cycleways,"* and *"Mass Rapid Transit."* Each strategy was then evaluated collectively using dot voting, with colored dots representing different criteria: equity/accessibility (pink), livability/well-being (yellow), and cost-efficiency/value for money (black).

This participatory process helped to identify which strategies should be prioritized based on their impact, feasibility, and alignment with the group's values.

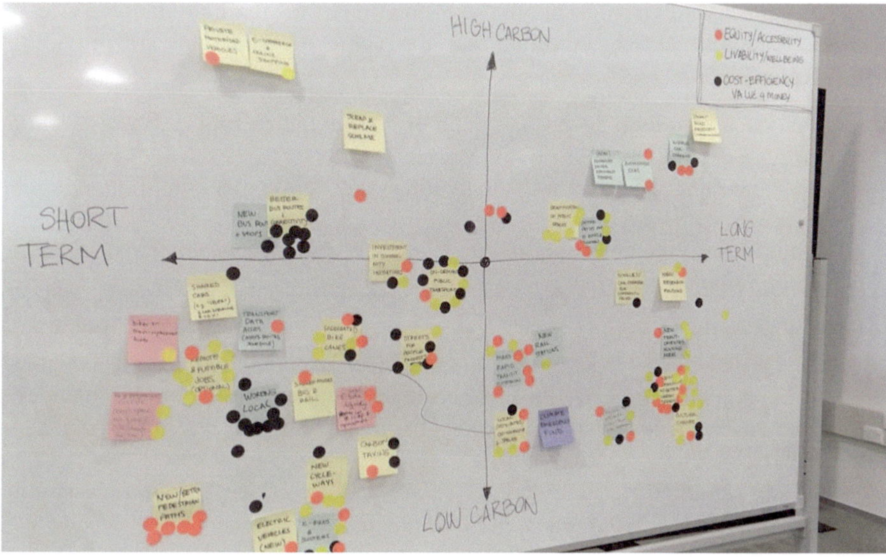

Fig. 8.7 Snapshot of "Decarbonising Transport" Workshop organized by the Transportation Research Centre at the University of Auckland (2022)

Idea Visualisation

Difficulty: Medium
Duration: 1 h.
Materials

- Pens
- Template printed in A4 size—or no template at all

Template link

An image speaks a thousand words. The visualization tool helps teams create visual representations of their project, service, or product, usually called ***Hero Shots***. These visuals help us communicate the project's value and context of use in a clear and relatable way, allowing stakeholders, users, or potential partners to picture themselves engaging with and benefiting from the offering.

How It Works

The tool suggests that teams create two types of hero shots.

1. Hero Shot of the Product or Service Itself

This is a styled visual or photo representation of the main product, service, or offer. It should clearly communicate the value of what you are providing. Examples include:

- A physical product, such as a piece of equipment or tool. E.g.: *sustainable camping gear.*
- A digital interface, like a website landing page or app mock-up. E.g.: *a digital interface of an educational game.*
- A snapshot of the key value of an experience. E.g.: *the menu of a sustainable seafood dining experience.*

2. Hero Shot of the project in the Context of Use

   This visual places your product, service, or experience within its intended environment, showing extra info like scale or intended use. Examples include:
   - A person using sustainable camping gear in a forest. E.g. *collecting rainwater with it.*
   - Kids using an educational game to learn about conservation in a classroom. E.g.: *interaction with the teacher and the digital platform.*
   - A symbolic representation of the impact of your project. E.g.: *an image of sustainable fishing communities and thriving marine ecosystems as a consequence of your intervention*

Steps:

1. Define Your Offer: Identify the product, service, or value you want to communicate. What do you want people to understand or feel when they see the image?
2. Visualize Impact: Brainstorm the most effective way to represent the offer and its use. Focus on clarity, relatability, and emotional connection (but keep it real!).
3. Capture or Design the Shots: Create the two visuals—one showing the product/service and the other showcasing it in action or delivering its value in context.

By creating these two types of hero shots, teams can effectively communicate their project's value, helping audiences not only understand the offer but also see its tangible impact in real-world contexts.

If your team doesn't have drawing or design expertise, AI tools can be a fantastic way to craft high-quality hero shots quickly and effectively. E.g.: By describing your educational game about conservation in detail, AI can generate an image of children engaging with it in a classroom.

Example Use

SeaStar: Reimagining The Seafood Restaurant Experience by Clara Schröder (2022).

SeaStar is a new certification that recognizes and awards Auckland seafood restaurants that demonstrate an exceptional commitment to sustainability. The SeaStar rating scheme encompasses five categories: Source, Treasure, Foster, Celebrate, and

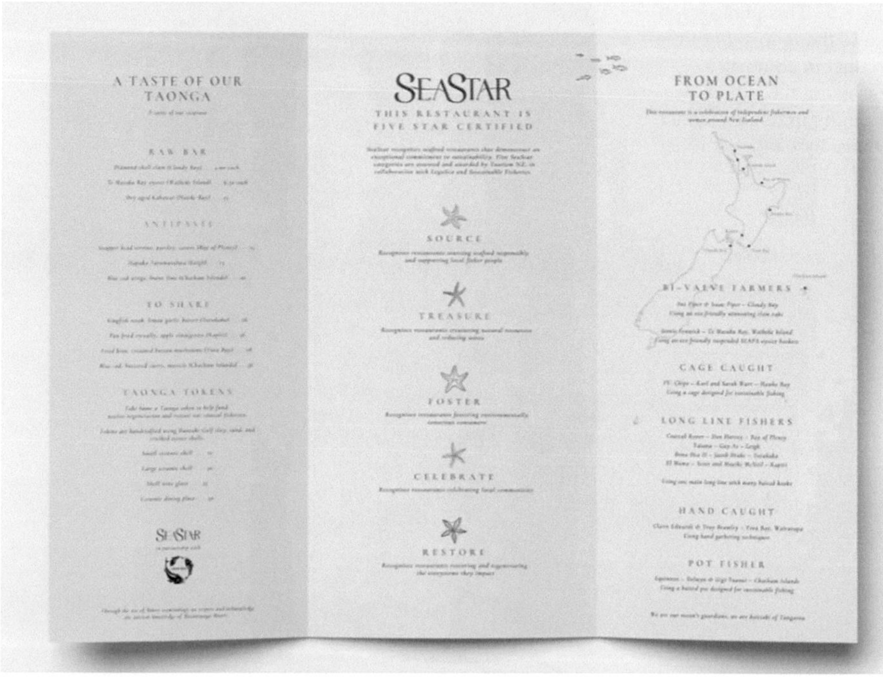

Fig. 8.8 This project is represented by the menu of a restaurant that has obtained the Seastar certification. Even though the project is much more than the menu, the menu makes the project tangible and it is an effective communication tool. (Image by Clara Schröder 2022)

Restore. Each SeaStar category is assessed and awarded by Tourism New Zealand in collaboration with LegaSea and Sustainable Fisheries.

This project reimagines the dining experience at Tāmaki Makaurau's seafood restaurants with the aim of making eco-conscious consumerism accessible and promoting restorative marine eco-tourism. This is a systemic initiative with multiple value points, and a decision was made to communicate it using the menu as the hero shot (Figs. 8.8 and 8.9).

Actionable Feedback Canvas

Difficulty: Medium
Duration: 30 min.
Materials

- Pens
- Template printed in A3 or A4 size

Fig. 8.9 This photograph shows the *hero shot* of the product in context, conveying the experience of dining in the restaurant. (Image by Clara Schröder 2022)

Template link

The Actionable Feedback tool is good for teams presenting their ideas to key stakeholders, potential funders, and community members. This tool helps structure feedback in a way that makes it clear, actionable, and easy to prioritize for further project development. By using the "4Ls" method[2]—focusing on what the audience Liked, Learned, Lacked, and Longed for—teams can organize feedback thoughtfully, capturing valuable insights about what resonated and identifying areas to refine or expand. By providing the audience with an agreed-upon framework, critical feedback will feel more comfortable for all parties involved, with no hurt feelings.

This feedback process is an important validation step. By listening carefully and organizing responses, the team can improve their project and build trust with their audience, demonstrating that they're open to learning and adapting.

---

[2] The 4Ls method was initially developed by Mary Gorman and Ellen Gottesdiener for EBG consulting.

How It Works

1. Set up: Communicate how the 4Ls work to the audience, you can give them the key attached to the template. Also, designate one team member to capture feedback. This ensures nothing gets missed while other team members engage with the audience.
2. Categorize Feedback Using the 4Ls:

   – Liked: What did the audience enjoy or appreciate about your presentation? This could be a specific feature or a general approach.
   – Learned: Did you gain new information or perspective from the feedback? This category includes insights or suggestions that could enhance your project.
   – Lacked: What did the presentation miss? Was something unclear or underdeveloped?
   – Longed for: What new elements did the audience wish for? This could be ideas they'd like to see integrated to make the solution even stronger.

3. Create Action Points: After the pitch, review and organize the feedback into actionable tasks. Outline what needs improvement, identify next steps, and assign responsibilities.

Using this tool, feedback becomes a guide for refining your project, ensuring that the team continuously improves and aligns with stakeholder expectations (Fig. 8.10).

Example Use

This tool helps teams to iterate by capturing feedback in real-time and breaking it down into actionable tasks, seamlessly assigning responsibilities to team members (Fig. 8.11).

Service Delivery Blueprint

Difficulty: Medium
Duration: 2 h.
Materials

– Pens, sticky notes
– Template printed in A0 or A1 size—one per team

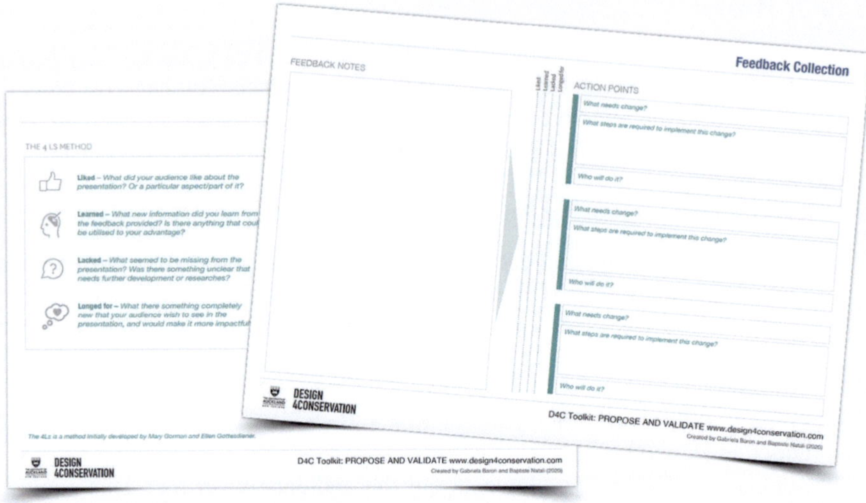

Fig. 8.10 The feedback collection template. (Gabriela Baron 2021)

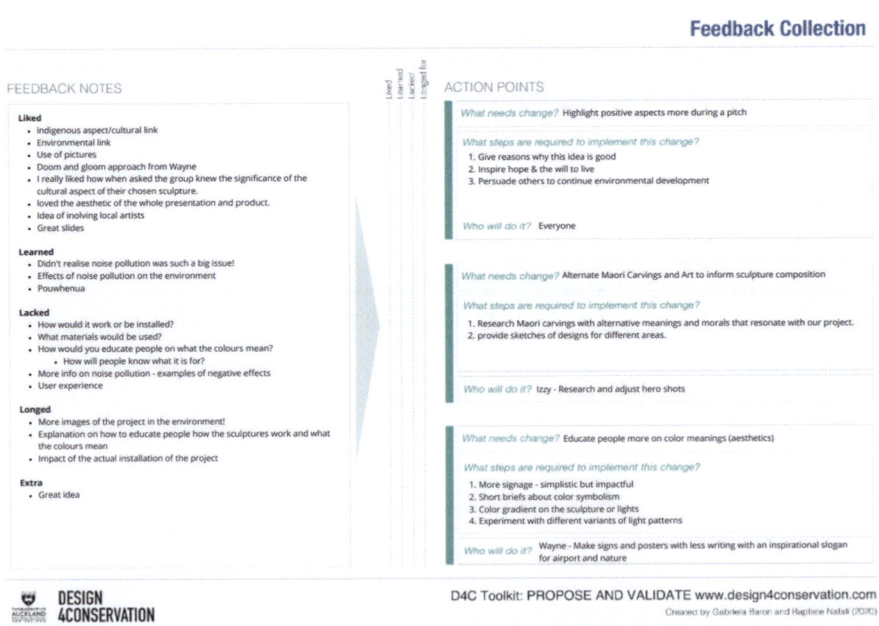

Fig. 8.11 The feedback and iteration tool used by students for The Atmos project. Authors: Siena Thompson, Isobel Mullins, Isobel Caldwell, and Wayne Poon (2022)

Template link

The Service Delivery Blueprint helps teams fine-tune their project by mapping out how people interact with the service or solution they're developing. This canvas is best used once your project idea has been validated by key stakeholders, allowing you to work out the practical details of delivering it effectively.

Service blueprinting is a valuable tool for designing and delivering seamless, user-centered services by providing a detailed visualization of the entire service process. By mapping out both the "frontstage" (visible to users) and "backstage" (behind-the-scenes) activities, blueprinting helps teams understand every component needed to create a positive user experience.

Here's why you'll love service blueprints:

- Clarity and Coordination: Blueprinting breaks down complex service interactions into organized steps, allowing each team member to understand their role and how it connects to others. This structured approach reduces miscommunication and ensures everyone is aligned on the service goals and processes.
- User Experience: By visualizing the user journey in detail, teams can identify moments where the user might feel frustrated, confused, or disconnected. This lets teams proactively improve the user experience, ensuring every touchpoint is positive and intuitive.
- Operational Efficiency: Service blueprints reveal the connections between frontstage and backstage operations, making it easier to spot inefficiencies, redundancies, or bottlenecks. Addressing these areas improves the service flow and makes resource use more effective.
- Framework for Evaluation and Iteration: Blueprinting gives a clear structure for assessing and refining the service over time. Teams can continually update the blueprint based on user feedback or changing needs, making it a dynamic tool for long-term improvement.

How It Works

1. Set Global Goals (optional): Identify which Sustainable Development Goal(s) your project aligns with, circling the relevant goals to keep your broader mission in focus.
2. Define Personas: Choose your main human persona (e.g., the primary user) and add details like their name and key characteristics. If possible, run this activity for several personas to get a well-rounded view.
3. Map the User Journey: Organize the "People" line along a timeline—what happens Before, During, and After they interact with your project? This helps outline the user experience step-by-step.
4. Outline Frontstage Activities: These are visible to users, including staff activities and interaction points (touchpoints) with the service. Touchpoints in a service blueprint are the moments when a customer interacts with a service. Think of them as specific points of contact that shape the customer experience. These can

occur in various ways, such as through in-person interactions, phone calls, emails, or even social media messages.
5. Identify Backstage Processes: These are the behind-the-scenes elements that support the user experience, like any necessary resources, support activities, and backend processes.
6. List Development Needs: Note anything additional you'll need to develop or adjust to make your project work smoothly, such as specific tools, resources, or training for staff.
7. Capture the User Experience: Think about how users might feel at each step—use emojis to mark high points and low points in the experience. This will help identify areas where the user journey can be improved.

Finally, check back to see if your plan effectively addresses the core conservation goals of your project. Use this process to keep refining your approach, focusing on creating a smooth and memorable experience for the people who engage with your project (Fig. 8.12).

Example Use

SeaGardens is a marine regeneration project on Auckland's waterfront, dedicated to restoring biodiversity in the Hauraki Gulf through artificial reef structures that support mussel reefs. By integrating interactive sculptures, educational displays, and partnerships with Māori artists, the project reconnects people with the harbor while enhancing water quality, reviving ecosystems, and promoting sustainable harvesting. This service delivery blueprint maps out the user journey and implementation process, illustrating how an Auckland CBD office worker, Jimmy Taylor, engages with the initiative (Fig. 8.13).

Let's analyze the key elements of this template:

1. User Experience Timeline
    - Initial Engagement: The user first sees the sculpture from their office window or on social media and gets curious.
    - Exploration: They walk along the waterfront, notice the glowing mussel sculpture, and observe the interactive info boards.
    - Deeper Interaction: They take photos, post on social media, and begin to understand the impact of mussel reefs on water quality.
    - Action & Advocacy: They join an action group, participate in citizen science, and tell others about the initiative.
    - Long-Term Impact: Over time, they notice cleaner water, more marine life, and an active community around SeaGardens.
2. Front-Stage Activities (What Users See)
    - Installation of "Te Ākau" sculpture, a physical landmark for the project.

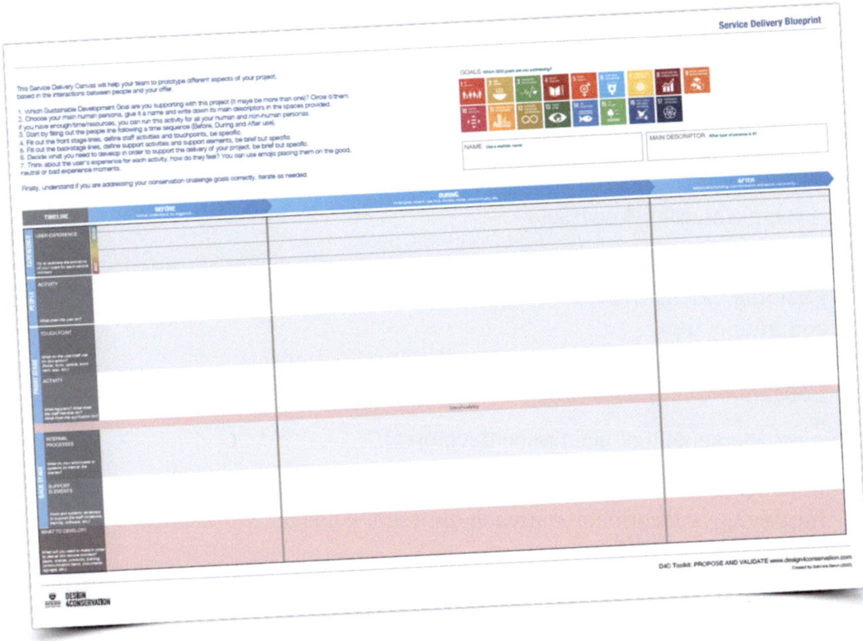

Fig. 8.12 The service delivery blueprint

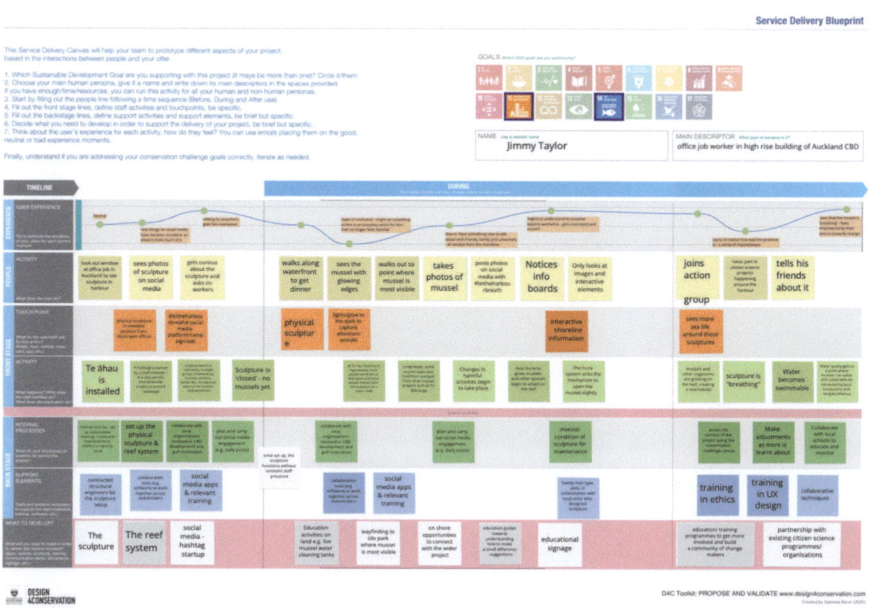

Fig. 8.13 The service delivery blueprint used by students for the Sea Gardens project by Cara-Sophia Andres, Elouise Kidd, Benjamin Jarvis, and Jade Nolan (2023)

- Mussel reef restoration with visual changes over time (e.g., mussels growing, kelp forming, water clarity improving).
- Interactive shoreline information to educate the public about the ecosystem and its recovery.
- The sculpture's movement ("breathing") as mussels grow, creating a dynamic visual indicator of ecosystem health.

3. Backstage (Internal Processes)

   - Setting up the reef system and collaborating with scientists, artists, and iwi.
   - Planning social media engagement and UX design for effective public communication.
   - Monitoring sculpture and reef health to assess impact.
   - Educational programs and partnerships with citizen science initiatives.

4. Support Elements that need to be developed:

   - Enhancing the sculpture and reef system with better designs.
   - Improving educational outreach through signage, workshops, and citizen science.
   - Creating wayfinding tools to help people locate and engage with mussel growth areas.

- Training in ethics and UX design to ensure inclusivity and effectiveness.

User Experience
The line at the top of the blueprint maps a transformative journey where an everyday office worker gradually connects with marine conservation, from passive observation to active participation. The initiative fosters a sense of ownership over Auckland's waterfront, using design, science, and storytelling to regenerate marine life and engage the community.

Chapter 9
Plan for Impact

Gabriela Nuri Baron

When designing for the natural environment, it is crucial to verify that your project is making a real, positive difference. Planning for impact before deployment allows you to set aside resources for measuring this impact, like time and money. This step is crucial because it will not only boost your project's credibility but also give you a clear view of whether your efforts will actually help regenerate and restore the natural environment.

Measuring impact can sometimes feel like a big task, especially in environmental projects where there are so many factors to consider. That's where the Agile-to-Deep approach comes in. With the Agile-to-Deep approach, you don't need to wait until you have a huge budget or a large team to start measuring your impact. Instead, you can begin with lightweight, quick methods like simple surveys, observation logs, or basic metrics, that give you a snapshot of your progress. Over time, as your project grows or gains resources, you can dive deeper, using more detailed measurement techniques to capture a fuller picture of your impact.

G. N. Baron (✉)
Faculty of Engineering and Design, University of Auckland – Waipapa Taumata Rau, Auckland, New Zealand
e-mail: g.baron@auckland.ac.nz

G. N. Baron (ed.), *Design for Environmental Conservation*,
https://doi.org/10.1007/978-981-96-9574-4_9

Sustainability Radar and Checklist

Difficulty: Medium
Duration: 1 h.
Materials

- Pens
- Template printed in A3 size—one per team

Template link

The D4C Sustainability Radar serves as a powerful tool for refining a project's alignment with the Sustainable Development Goals (SDGs) by summarizing best practices for environmental sustainability on one side and social sustainability on the other. On the environmental front, it highlights strategies such as extending system lifespan, using renewable materials and energy, minimizing waste, and reducing transport impacts, helping projects lower their ecological footprint. On the social side, it promotes ethical labor conditions, responsible consumption, and community regeneration, ensuring that projects contribute positively to society. By evaluating a project through this lens, teams can identify areas for improvement, integrate sustainability more holistically, and create solutions that balance ecological integrity with social equity.

Think of it as your roadmap for brainstorming concrete strategies to make your initiative better for the planet and the community.[1] It helps teams create a visual summary of their sustainability initiatives, allowing you to see where you stand now and where you want to go. By identifying "hotspots" (areas where you can make significant improvements) your team can focus your efforts effectively.

How It Works

1. Review the Strategy Checklist: Understand if the strategies from the checklist are applicable to your project: Define the priority level of each strategy in relation to your goals/values/visions. Note that sometimes these strategies are not relevant, are hard to implement, or fall outside the scope of the project (this is ok as long as you are aware of it).
2. For each strategy, assign a priority level:

   L for low-priority
   M for medium-priority
   H for high-priority
   N for not applicable

[1] This template can also be used in the propose and Validate Phase of the D4C methodology, as the checklist provides valuable inspiration during ideation sessions.

Write down each strategy's priority on the corresponding axis of the Radar Canvas. This creates a visual representation of where your project stands.

3. Document Your Actions: In the text boxes on the canvas, describe the actions your project is currently taking related to each strategy. Briefly explain how these actions contribute to your sustainability goals.
4. Evaluate Your Progress: Now, assess how well your project is addressing each strategy. Use a scale:

   – worsened: no progress
   = same status: maintained current efforts
   + improved: some progress made
   ++ really improved: significant improvements

   Mark a point in the radar intersection based on your evaluation.

5. Connect the Dots: Draw a polygon by joining the points you marked on the Radar. This shape will give you a visual summary of your project's strengths and areas needing attention.
6. Conduct a Visual Analysis: Finally, analyze the shape of your polygon. Look at how your contributions align with the importance of each strategy. Identify critical hotspots—these are strategies marked as high or medium priority. Make sure you're actively working to improve these areas.

As you work through this process, it's important to keep refining your strategies. You don't need to check every box —focus on leveraging your project's strengths and addressing its biggest challenges. This tool is about making sure that your idea is as sustainable as possible and creating a visual summary of where you stand in terms of environment and community (Fig. 9.1).

Example Use

The checklist section outlines several key initiatives to enhance environmental and social sustainability, providing a structured approach for refining projects to align with the Sustainable Development Goals (SDGs) (Fig. 9.2).

Environmental Sustainability

1. Servicisation: Encourages shifting from product ownership to shared use models like rental or leasing to minimize material and energy consumption.
2. Use of Renewable Materials and Energy: Promotes local, renewable, recycled, or biodegradable materials, along with carbon offsetting and renewable energy adoption to reduce the project's environmental impact.
3. Transport and Packaging Reduction: Suggests digital solutions to minimize physical transport and packaging, reducing emissions and resource consumption.

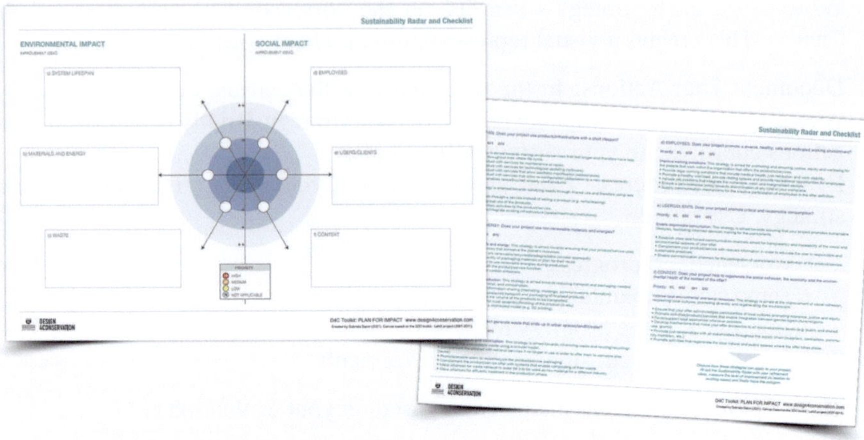

Fig. 9.1 The sustainability radar and checklist template. (Created by Gabriela Baron in 2020. This Canvas is based on the SDO toolkit—LeNS project (2007–2011))

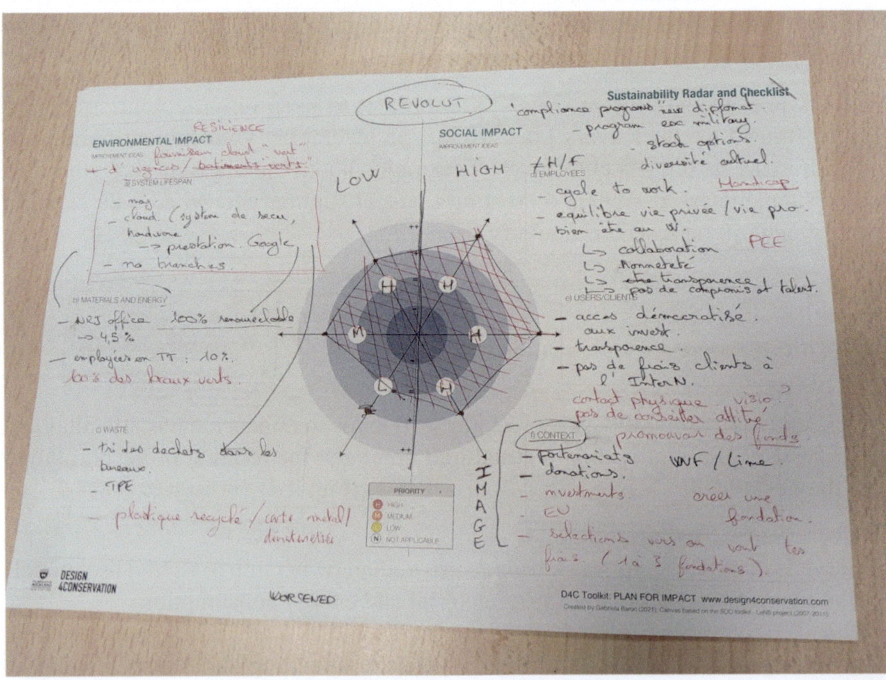

Fig. 9.2 Analysis of an existing business solution using the Sustainability Radar. In this image, you can see how this Canvas is used for brainstorming and is always a work in progress. Taken during a Postgraduate Business Seminar for IAE Lyon, France, 2025

4. Waste Minimization & Valorization: Encourages reuse, recycling, and composting while integrating circular economy strategies such as industrial symbiosis (where one industry's waste becomes another's resource).

Social Sustainability

1. Improving Working Conditions: Focuses on fair wages, safe environments, and inclusivity in employment, ensuring justice, equity, and well-being for all workers.
2. Enabling Responsible Consumption: Encourages transparency and traceability, educating consumers on sustainable choices and promoting co-creation with users.
3. Context-Based Regeneration: Supports local economies, cultural diversity, and ecosystem restoration, ensuring that conservation projects respect and enhance the communities they serve.

By applying and reflecting on these strategies, projects can balance ecological responsibility with social equity, creating sustainable, inclusive solutions that contribute meaningfully to global conservation efforts.

Impact Planning Canvas

Difficulty: Medium
Duration: 1 h.
Materials

- Pens
- Template printed in A3 size—one per team

Template link

Using the Impact Planning tool helps you think critically about the full scope of your project and prepares you for both intended and unintended consequences. By identifying what you want to achieve and how to measure it, you create a roadmap for success that not only tracks progress but also ensures that you're mindful of the broader impact of your work (Fig. 9.3).

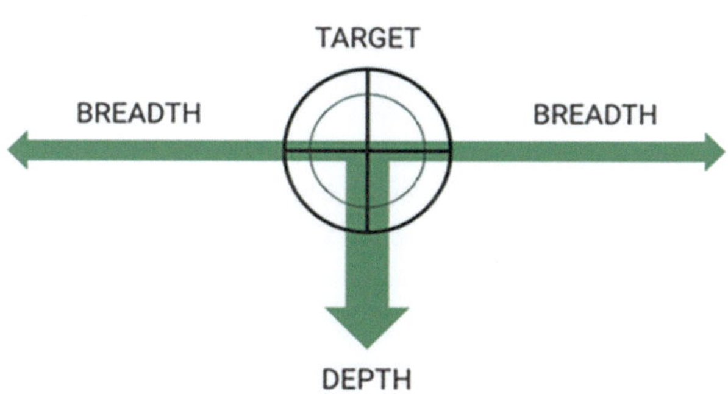

Fig. 9.3 The impact planning canvas helps explore indicators for breadth, depth, and target of your impact, while also keeping data lean. (Gabriela Baron 2021)

How It Works

1. Define Your Goals: Start by writing down the main goals of your project. Discuss with your team how you will know if you are making the intended changes.
2. Plan for Intended Impacts:

   Questions: With your team, discuss the breadth (how wide-ranging your impact will be), depth (how profound the impact will be), and target (how aligned is this indicator the the project's ultimate objective) of the impact you'd like to have. Start by asking the right questions: What will you want to find out? What questions will need to be answered? For instance, if your project aims to improve local biodiversity, consider how many species you want to impact, the ecosystems involved, and the communities that might benefit.
   Indicators: What indicators will you use to measure success? For example, if your goal is to increase community engagement, you might measure attendance at events or the number of participants in activities. In general, breadth is measured with quantitative indicators (numbers), while depth tends to be more qualitative (but not always).
   Tools: You will need some tools to measure these impacts, what is a realistic tool that you will be able to use to respond to your impact questions? These should be viable economically, so consider resource constraints such as time and funding. These tools may be surveys, an app, interviews, field observations, etc.

3. Plan for unintended impacts:

   Consider the Systemic Nature of Your Environment by remembering that your project exists within a larger system, and your actions might influence other systems in unexpected ways.
   Unintended impacts are the effects of your project that you may not have foreseen. It's okay if you don't know everything about these impacts right now; the key is to acknowledge that they will exist.
   Use the brainstorming template to explore these unintended impacts. Discuss possible unintended impacts with your team, and organize them into categories. How will you keep an eye on these impacts?
   For example, if your project involves planting new trees, a potential unintended impact could be an increase in unwanted pest species, which might affect the balance of endemic species in the area. Consider how this could influence other environmental factors and what you might do to monitor those changes, and take adaptive action if necessary.
   As you work through these steps, keep in mind that impact planning is an ongoing process. Regularly revisit your goals and the impacts you're measuring to adapt as needed. This reflective practice helps ensure that your project continues to evolve and make a meaningful difference in the community and the environment (Fig. 9.4).

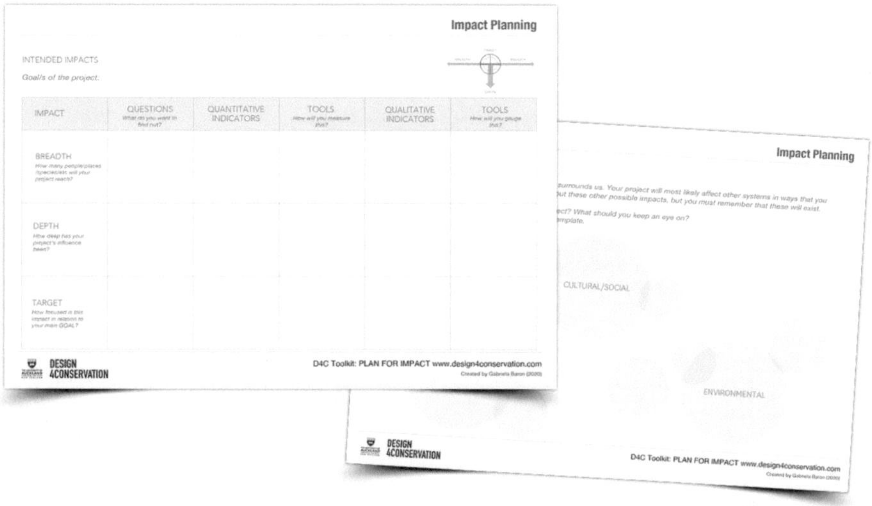

Fig. 9.4 The impact planning canvas. (Gabriela Baron 2021)

Example Use

The Sea Gardens project, which focuses on marine regeneration in Auckland's waterfront, used this framework to evaluate and refine its engagement strategies. Here's how (Fig. 9.5):

1. Breadth: Expanding Reach & Engagement

   - Key Question: Will the project engage a diverse range of the public and encourage participation?
   - Indicators: Measuring social media interactions, waterfront user surveys, and monitoring public engagement in citizen science activities.
   - Application:

   – Tracking participation in interactive sculptures and educational displays.
   – Ensuring that a broad range of people (residents, tourists, school groups) interact with the project.

2. Depth: Changing Perceptions & Behaviors

   - Key Question: Will people perceive Waitematā Harbour as a living system in need of care?
   - Indicators:

   – Observing whether users see, interact with, and understand ecological displays.
   – Noting positive comments and increased involvement in volunteer activities.

   - Application:

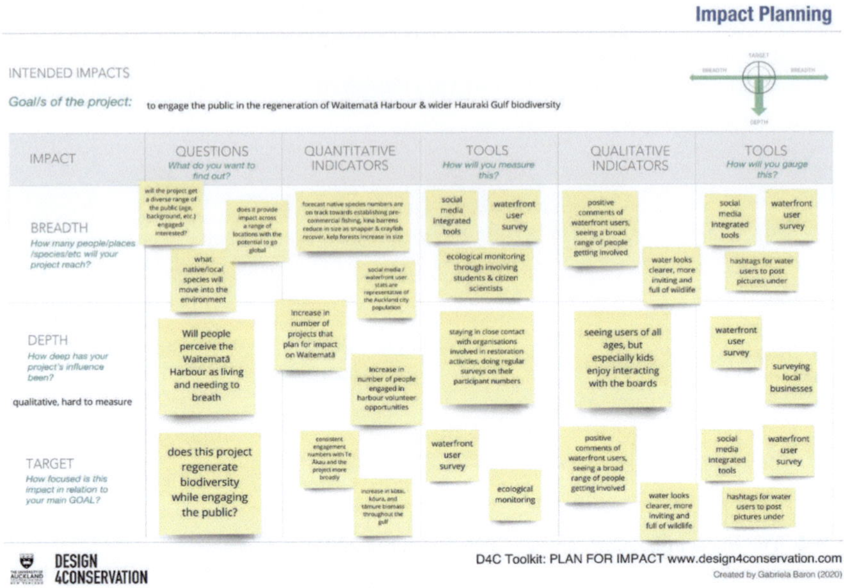

Fig. 9.5 The impact planning canvas used by students for the Sea Gardens project by Cara Andres, Elouise Kidd, Benjamin Jarvis, and Jade Nolan, 2023

- Assessing whether visitors feel more connected to the marine ecosystem.
- Measuring changes in local conservation behaviors, like reduced littering or interest in sustainable fishing.

3. Target: Achieving Biodiversity & Engagement Goals

- Key Question: Does this project regenerate biodiversity while engaging the public?
- Indicators:

- Growth of mussel reef structures, clearer water, and increased marine life.
- Long-term participation in conservation activities.

- Application:

- Conducting ecological monitoring to track biodiversity improvements.
- Partnering with Māori artists and scientists to reinforce cultural and ecological restoration efforts.

By using this framework, the Sea Gardens team can ensure that their project effectively reaches people, deepens awareness, and delivers tangible ecological benefits. The structure also helps in measuring success, guiding iterative improvements, and making a strong case for funding and policy support (Fig. 9.6).

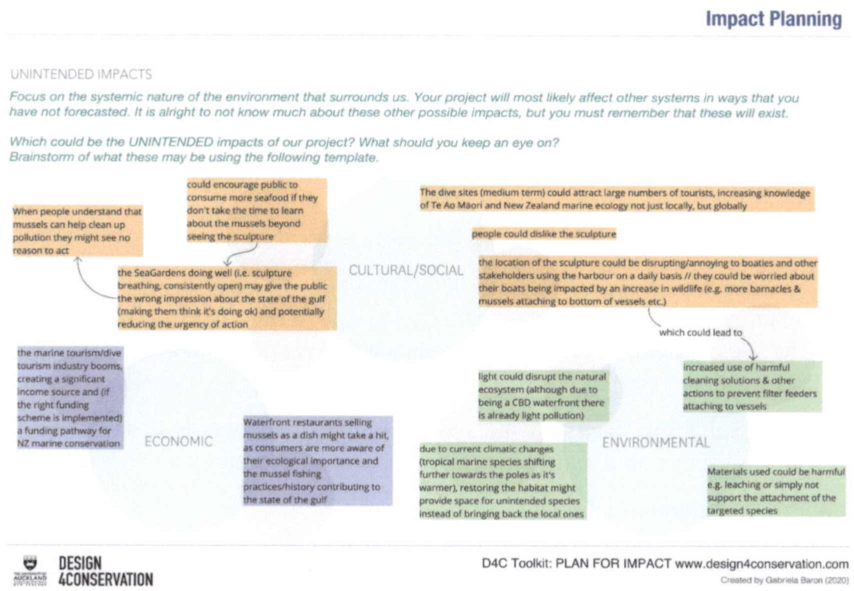

Fig. 9.6 Unintended Impacts analysis for the Sea Gardens project by Cara Andres, Elouise Kidd, Benjamin Jarvis, and Jade Nolan, 2023

While the SeaGardens project aims to restore biodiversity and engage the public in marine conservation, it may also lead to unintended consequences across cultural, environmental, and economic dimensions.

Cultural & Social Unintended Impacts

- Increased Seafood Consumption: Without proper educational framing, public exposure to mussel reefs could encourage more people to eat mussels, rather than understanding their ecological role.
- False Sense of Environmental Health: If the sculpture and reef structures appear thriving, the public may assume that the Hauraki Gulf is recovering well, potentially reducing urgency for broader conservation actions.
- Complacency in Pollution Reduction: Learning that mussels help filter pollution could lead to the perception that human actions are less necessary for maintaining water quality.
- Tourism Pressures: If dive sites around the sculptures attract large numbers of tourists, this could bring positive cultural exchange but also increased environmental strain.
- Community Acceptance Issues:
  - Some people may dislike the aesthetic of the sculpture.
  - Boaties and harbor users might be frustrated by the structure's location, worrying about navigational disruption or increased marine life fouling their vessels.

Environmental Unintended Impacts

- Species Displacement: Due to climate change, warmer waters could encourage the arrival of tropical species, possibly outcompeting native species the project aims to restore.
- Light Pollution Effects: Though the CBD waterfront already has high artificial lighting, additional illumination from the sculpture could further disrupt marine ecosystems.
- Marine Biofouling & Anti-Fouling Measures:
  - Increased mussel and barnacle attachment could lead to a rise in harmful cleaning practices, as boat owners try to remove filter feeders from hulls.
  - The use of chemical anti-fouling agents could introduce toxins into the water, counteracting conservation efforts.
- Material Suitability Risks: If the sculpture materials degrade or leach harmful substances, they could damage rather than support the targeted ecosystem.

Economic Unintended Impacts

- Marine & Dive Tourism Growth: If successful, the project could boost eco-tourism, potentially creating a funding stream for marine conservation—but only if a sustainable funding model is in place.
- Impact on Seafood Industry: Increased public awareness of mussel reef destruction and ecological importance might reduce demand for farmed or wild-harvested mussels, affecting local seafood businesses.

Mitigation Strategies: To address these risks, SeaGardens could:

- Ensure educational messaging clarifies that mussels are ecosystem engineers, not just seafood.
- Balance storytelling so that the public understands restoration efforts are ongoing, and the Gulf still needs protection.
- Work with boating communities to minimize disruption and address concerns about marine fouling.
- Use eco-friendly materials and monitor for unintended species shifts.
  Establish a responsible marine tourism model, ensuring conservation remains the priority.

Advanced Impact Planning

Difficulty: High
Duration: 2 h.
Materials

- Pens
- Template printed in A3 size—one per team

Template link

The advanced impact planning canvas allows teams to refine and add detail to the intended impacts canvas. The tool is structured around key elements that allow you to document and monitor the intended impacts of your project over different timeframes. This advanced canvas includes sections for setting outcomes, indicators, data types, measurement tools, and aligning with global metrics like the UN Sustainable Development Goals (SDGs). This tool works well alongside ***Theories of change***[2] or can be used after refining your initial intended impacts canvas.

How It Works

1. Project Name and Timeframe:
   - Begin by entering your project name at the top of the form.
   - Define the timeframe for your impact measurements. You can set goals for Short Term, Medium Term, and Long Term to show of progress over time.
2. Define Impact Categories and Outcomes:
   In the Impact Category column, list the areas where you want to see change (e.g., environmental, social, economic). For each category, write down specific outcomes you hope to achieve. These outcomes represent the changes or improvements you're aiming for.
3. Set Outcome Indicators: In the Indicators column, define specific metrics you'll use to track each outcome. These should be quantifiable or describable aspects that provide evidence of impact (e.g., number of trees planted, reduction in waste, community engagement levels).
4. Determine Units and Data Types: Specify the Units for each indicator, which could be numbers, percentages, or other relevant metrics. Understand whether your data will be Quantitative (numerical) or Qualitative (descriptive) and note this in the UNITS column.
5. Define Measurement Tools: In the Tools column, outline how you'll gather data for each indicator. Examples might include surveys, field observations, or scientific measurements. This section helps you plan the methods and tools you'll need to collect reliable data.
6. Align with UN SDGs (Optional): If your project contributes to specific UN Sustainable Development Goals (SDGs), list the relevant SDG metric(s).[3] This alignment helps contextualize your project's impact in a broader, globally recognized framework.

---

[2] A Theory of Change (ToC) is a comprehensive framework that outlines how and why a desired change is expected to happen in a specific context. It identifies the relationships between activities, outputs, outcomes, and long-term impact, helping to map the pathway to achieving a goal.

[3] UN SDG metrics are a set of indicators used to measure progress toward the United Nations Sustainable Development Goals (SDGs). These metrics help track global, national, and local efforts in areas such as poverty reduction, education, health, climate action, and economic growth, ensuring accountability and data-driven decision-making.

Fig. 9.7 The advanced Impact Planning Canvas. (Developed by Gabriela Baron 2022)

7. Record your Baseline Metric: it's very important to represent your starting point before any interventions. If you don't have initial data, consider the first data collected as your baseline. The baseline will give you a reference to assess how much change occurs over time (Fig. 9.7).

Example Use

Suppose you're running a community tree-planting initiative. Here's how you might fill out each section:

- Example Intended Impact: Enhance urban biodiversity, improve air quality, and foster community engagement in conservation efforts.
- Impact Category: Environmental
- Outcome: Increased urban greenery
- Indicator: Number of trees planted per year
- Units: Trees planted
- Data Type and Parameters: Quantitative, counting each tree
- Tools: On-site counts, GPS mapping
- UN SDG Metric: Goal 15 (Life on Land), Indicator 15.1.1 (Forest area as a proportion of total land area)
- Baseline Metric: Current number of trees in the area

Chapter 10
Deploy Sustainably

Gabriela Nuri Baron

This phase is all about bringing your project to life. After all the planning, testing, and refining, it's time to deploy your idea in the real world! Whether your project is for-profit or nonprofit, this phase helps you get started on a smaller, local scale before expanding to a wider audience. By starting locally, sustainably, and forging genuine partnerships, you're creating a strong foundation that allows your project to grow in a way that's impactful and lasting.

Whether your project aims to generate profit or purely serve a social/environmental mission, sustainable business models can help guide your growth. This means planning ways for the project to support itself over time and not lose its alignment with impact. In a for-profit model, this might mean developing products or services that align with your mission, while in a nonprofit model, it could involve creative funding strategies or partnerships that keep the project going strong without compromising its values.

G. N. Baron (✉)
Faculty of Engineering and Design, University of Auckland – Waipapa Taumata Rau, Auckland, New Zealand
e-mail: g.baron@auckland.ac.nz

G. N. Baron (ed.), *Design for Environmental Conservation*,
https://doi.org/10.1007/978-981-96-9574-4_10

Contextual System Map

Difficulty: Medium
Duration: 1 h.
Materials:

- Pens, color markers
- Template printed in A3 or A2 size—one per team

Template link

The Contextual System Map is a highly visual tool that helps you understand and communicate the network of people, organizations, resources, and relationships that make your project work. It shows how different stakeholders come together to support your project, using highly visual elements like icons and keywords. This tool is also useful for project refinement in the "Propose and Validate" Phase, as it provides suggestions on how to make your project better.

There are 4 main elements to a system map:

- Stakeholders: people, organizations, or even non-human elements
- Places: physical or digital locations
- Things: support elements, products, and services
- Connections: between them (like information, labor, materials, or money flows)

How It Works

1. Create a Rough Draft on Paper:
   - Start with a blank sheet and place your project or *offer* in the center.
   - Around it, add all the stakeholders involved. Stakeholders could be:
   - People: Community members, project team members, funders, or customers.
   - Institutions: Schools, NGOs, businesses, or government agencies.
   - Non-human stakeholders: Animals, plants, or other elements of the ecosystem.
   - Draw arrows between your project and each stakeholder to show the types of relationships or "flows" connecting them. Flows could be labor, information, materials, energy, or money. Use different colors for each type of flow.
   - Add locations to show where these interactions take place if you feel these are relevant.
   - Be visual! Use icons and some keywords if necessary.
2. Organize Your Map on the Template: Once you're happy with your rough draft, transfer it to the provided template for a cleaner, more readable layout. Arrange everything neatly and make sure all the important information is included so that anyone looking at it will understand the system that supports your project.

3. Add a key to your system map, explaining the color code you have used for each flow, or any other key to reading your map.
4. Zoom In and Out as Needed:

   – You might find it helpful to create multiple versions of your map. One could provide a general overview of your project's context, while others could zoom into specific parts with more detail.

By mapping out the relationships, you'll be able to see any gaps, overlaps, or opportunities to strengthen your project's impact. Plus, it's a great visual tool to communicate your project's context to others! (Fig. 10.1).

Example Use

The icon tokens allowed participants to discuss partnerships, make agreements, and move things around. The two images above show the *draft* version and the *tidy* version of a System Map. But again, all the Canvases presented here are intended to be works in progress, as with any *iterative* process (Figs. 10.2, 10.3, and 10.4).

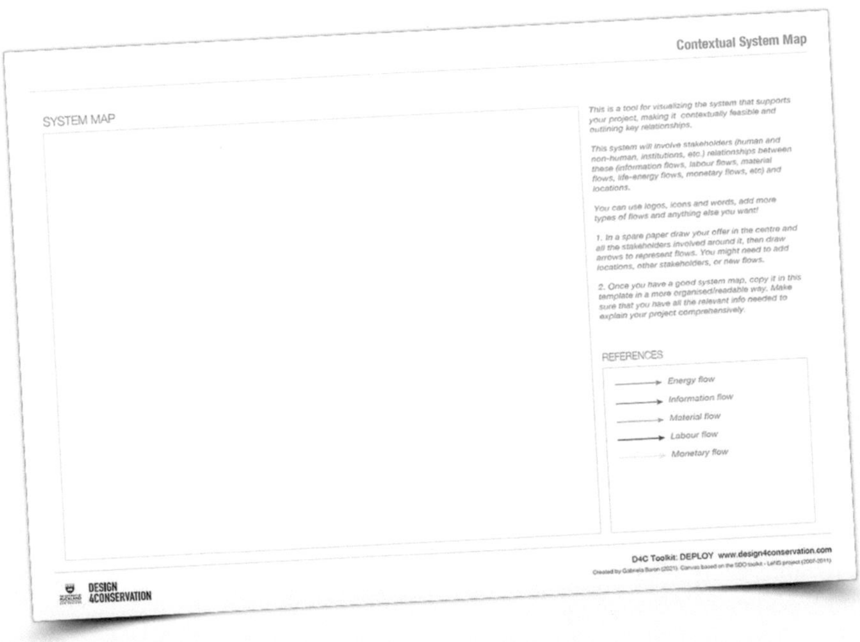

Fig. 10.1 The contextual system map canvas. (Developed by Gabriela Baron 2021)

Fig. 10.2 Student presenting the contextual system map for the "Funding in Love" project. (Lyon, France, 2025)

Stakeholder Motivation Matrix

Difficulty: Easy
Duration: 30 min.
Materials:

- Pens
- Template printed in A4 or A3 size—one per team

Template link

The Stakeholder Motivation Matrix is a tool that helps teams understand and visualize why different people and organizations (stakeholders) would want to be involved in a project. It helps stakeholders see where there are shared benefits, making it easier to create influence groups—groups of people and organizations with similar motivations who can work together effectively. By aligning these motivations, you're more likely to create lasting, impactful partnerships that make your project stronger and more sustainable.

Fig. 10.3 Initial version of a co-created system map, using pre-designed icons to make the exercise more playful. Wood tokens help participants move the icons around while discussing relationships within the system. (Project in partnership with Uru Whakaaro, Photo by Eric Jang 2022)

How It Works

1. List Stakeholders:
   - Start by writing the names of each (main) stakeholder in the gray boxes along both the top row and the first column of the matrix. Each stakeholder's name should appear twice, once on the top and once on the side. This chart allows you to look at each pair of stakeholders and see where their interests might overlap.
2. Identify Motivations:
   - For each stakeholder, write down their main motivations in the green boxes next to their names. Ask yourself: Why would this stakeholder be interested in participating in the project?
   - For example, if a local business is one of your stakeholders, their motivation might be to improve their public image or support the community. If a conservation group is involved, their motivation might be to protect local biodiversity.
3. Map Partnership Benefits:
   - Now, look at each pair of stakeholders and consider what benefits they might gain by working together on this project. In the intersecting boxes of each pair, write down specific benefits that this specific partnership would bring.

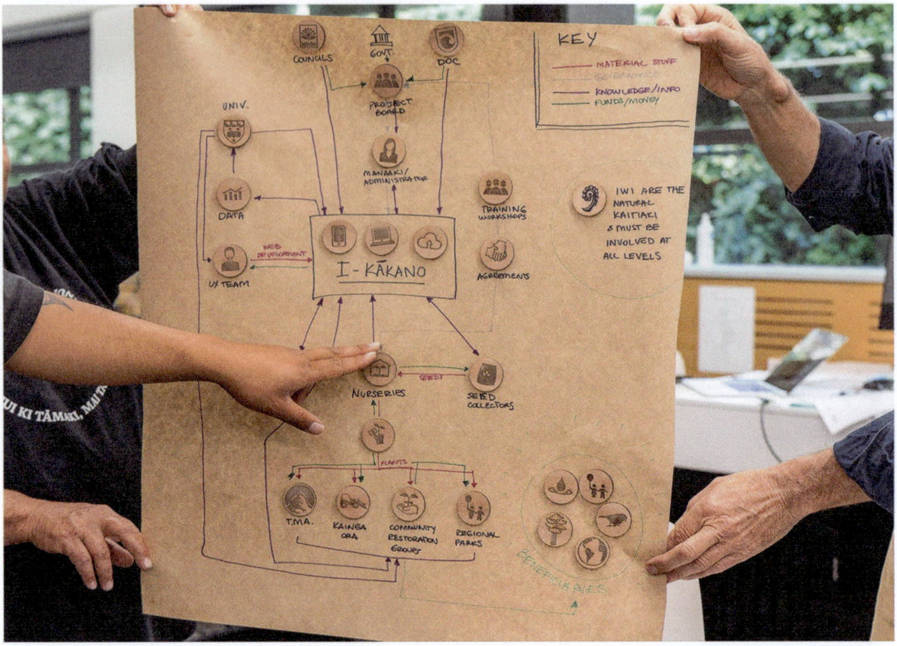

Fig. 10.4 Participant explaining how a project works using the a tidy version of a co-created system map, using pre-designed icons to make the exercise more playful. (Project in partnership with Uru Whakaaro, Photo by Eric Jang 2022)

- For example, if you have a local school and an environmental NGO, the intersection could say, *"Educational workshops on conservation."* This section highlights the opportunities for collaboration between stakeholders and shows how partnerships can help meet each group's goals.

The matrix can reveal unexpected overlaps, new partnership opportunities, and ways to align stakeholders' interests to create a cohesive, supportive network for your project, where everyone wins (Fig. 10.5).

Example Use

In this example, the SeaStar initiative presents a transformative vision for Auckland's seafood industry by integrating sustainability, consumer awareness, and marine conservation. By recognizing and awarding restaurants that demonstrate an exceptional commitment to sustainable practices, SeaStar not only fosters eco-conscious dining but also creates a bridge between the hospitality sector and marine restoration efforts (Fig. 10.6).

Fig. 10.5 The stakeholder motivation matrix. Developed by Gabriela Baron, 2021. This Canvas is based on the SDO toolkit—LeNS project (2007–2011)

STAKEHOLDERS	SeaStar	Seafood Restaurants	LegaSea	Local Artists	Tourism NZ	Tourists
SeaStar	Regenerate the marine environment, and enable conscious consumerism to foster marine ecotourism and restoration.	Seastar benefits by being able to enable conscious consumerism through restaurants and generate funds to regenerate the marine environment.	Seastar benefits by supporting initiatives regenerate the marine environment.	Seastar benefits from supporting the local artist community.	Seastar benefits by getting funding and promotion to carry out their mission of regenerating the marine environment.	Seastar benefit by engaging people in conscious consumerism and gaining funds to support marine restoration.
Seafood Restaurants	Seafood restaurants benefit through being sustainably certified to gain promotion from Tourism NZ and gain attention from the public to generate more customers and revenue. As well as supporting the local environment to ensure food security and mitigate the depletion of fish stocks.	Gain promotion from Tourism NZ and gain attention from the public to generate more customers and revenue. Support the local environment to ensure food security and mitigate the depletion of fish stocks.	Seafood restaurants benefit supporting the local environment to ensure food security and mitigate the depletion of fish stocks.	Seafood restaurants benefit by creating a unique and memorable dining experience with a token to take home.	Seafood restaurants benefit by getting international promotion	Seafood restaurants benefit by attracting more customers due to their Seastar certification
LegaSea	LegaSea gains awareness and funding for marine restoration projects.	LegaSea gains awareness and funding for marine restoration projects.	To ensure our lives are made richer through the health of our fisheries and the environment we all live in.	LegaSea gains by supporting the local artist community and getting promotion for their cause.	LegaSea gains awareness and funding for marine restoration projects.	LegaSea gains awareness and funding for marine restoration projects.
Local Artists	Local artists gain support and promotion for their products.	Local artists gain promotion for their work.	Local artists benefit by supporting the local marine environment and getting promotion for their products.	Get commissioned for projects that reflect their own sustainable values and allow for wide spread promotion of their work.	Local artists benefit by getting international promotion for their products.	Local artists gain a cut of Taonga Token revenue, and gain promotion for their work.
Tourism NZ	Tourism NZ gains by being able to promote a clean blue image of New Zealand sustainably certified restaurants.	Tourism NZ gains by being able to promote a clean blue image of New Zealand sustainably certified restaurants.	Tourism NZ gains by being able to promote a clean blue image of New Zealand's waters and give back to the environment.	Tourism NZ benefits by being able to ensure tourism gives back to our communities.	Committed to supporting initiatives that ensure tourism gives back to people, communities and the environment. Wish to foster a clean blue image of New Zealand to international tourists	Tourism NZ benefits by generating tourism revenue for New Zealand.
Tourists	Tourists benefit by gaining accessible knowledge about where seafood at restaurants comes from.	Tourists benefit by enjoying quality sustainable seafood	Tourists benefit by engaging restorative eco tourism to maintain food security.	Tourists benefit by getting a unique keep sake to take home with them and remember their travels to NZ seafood restaurants	Tourists benefit by learning about NZ and the renowned quality sustainable seafood.	Travel to NZ and enjoy the renowned quality sustainable seafood.

Fig. 10.6 Analysis of the SeaStar case study using the stakeholder motivation matrix. (Clara Schröder 2022)

Through a reimagined seafood restaurant experience, this initiative makes sustainable choices more accessible, encouraging diners to actively participate in the protection of New Zealand's marine ecosystems. By aligning the motivations of key stakeholders—restaurants, conservation groups, local artists, tourism organizations, and consumers—SeaStar cultivates a collaborative network that strengthens the link between food, culture, and environmental stewardship.

As restorative marine ecotourism gains momentum, SeaStar sets a precedent for how design and certification can drive positive change, ensuring that Auckland's seafood industry thrives without compromising the future of its ocean ecosystems.

Sustainable Growth Plan

Difficulty: Medium
Duration: 1 h.
Materials:

- Pens
- Template printed in A4 or A3 size - one per team

Template link

The Sustainable Growth Plan is a tool that helps you plan for the growth of your project in a way that keeps it impactful, financially stable, and environmentally sustainable. Growth here doesn't just mean becoming bigger—it's about reaching more people and ecosystems while staying true to your project's purpose. This tool is based on the Small, Local, Open, Connected (SLOC) model developed by Ezio Manzini.[1] The SLOC model encourages a distributed and decentralized approach to scaling, ensuring that your project grows in a way that respects local context, fosters community involvement, and connects globally to remain resilient.

SLOC Model Guideline:

- Small: Start on a local scale and grow gradually.
- Local: Stay grounded in the needs, desires, know-how, and resources of the community.
- Open: Keep the project adaptable and open to new ideas.
- Connected: Link your project to other communities, networks, and resources for support and resilience.

[1] Manzini, E. (2015). *Design, When Everybody Designs: An Introduction to Design for Social Innovation.* MIT Press. This book explores how design can support social innovation, with the SLOC model as a key framework for fostering sustainable and community-driven solutions.

How It Works

1. Pilot Project:

   – Start by describing the pilot phase of your project. This is the initial test or trial stage where you validate your concept, measure impact, and refine your approach.

2. Startup: Here, you're gathering feedback and learning what works in the real context over a longer time. This stage will include refinement as well and will serve as the foundation for your project's future growth.
3. Autonomy:

   – Next, consider how your project will become financially independent over time. Financial autonomy means having reliable revenue streams that allow your project to sustain itself without constantly needing outside funding. Many not for profit environmental projects start with an external grant, but then you need to identify potential revenue sources that align with your mission by securing local partnerships.

4. Replication:

   – Think about how your project could be replicated in other local contexts. Contextual replication means adapting your project to fit other communities, environments, or needs while keeping aligned with purpose and impact. What makes the project flexible and adaptable to different communities? How can you distribute the governance and financial resilience of each node?

5. Scaling:

   – When you're ready to grow further, develop a plan for resilient scaling. This involves expanding in a way that preserves the quality of your impact and maintains your connection to local communities. Consider if you need to scale at all. How can you ensure the project remains purposeful on a larger scale? (Fig. 10.7)

Example Use

Continuing with the SeaStar example, we can see how the SLOC Growth ethos can provide a growth roadmap (Fig. 10.8).

1. Pilot: Validate the concept through design students' research and prototyping.

   The pilot phase is a clear small-scale initiative, rooted in local expertise (design students and the university). It's an open process, allowing testing and refinement.

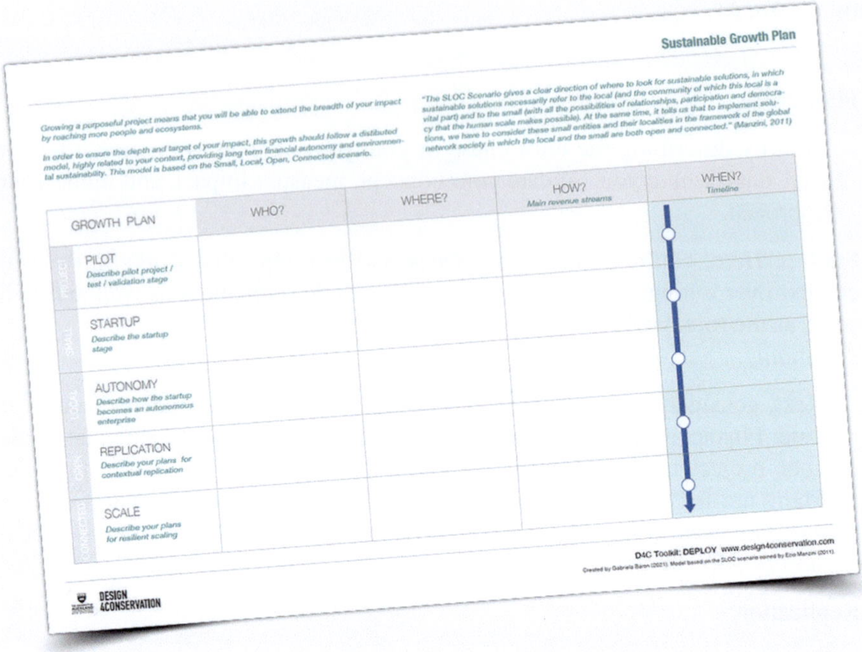

Fig. 10.7 The sustainable growth plan. (Developed by Gabriela Baron 2021)

GROWTH PLAN	WHO?	WHERE?	HOW? *Main revenue streams*	WHEN? *Timeline*
PROJECT — PILOT *Describe pilot project / test / validation stage*	Design 301 and 301 students validating concept	University of Auckland design studio for development testing	None, student volunteers	Immediately
SMALL — STARTUP *Describe the startup stage*	A sustainable seafood restaurant tests out the concept	A Sustainable Auckland viaduct seafood restaurant	University of Auckland velocity donation	1 Year
LOCAL — AUTONOMY *Describe how the startup becomes an autonomous enterprise*	Fisheries and restaurants choose to improve their practices and volunteer to be assessed to the SeaStar standard.	• Restaurants in New Zealand capital cities • Fisheries across New Zealand	Fish care and sustainable fisheries investment	4 Years
OPEN — REPLICATION *Describe your plans for contextual replication*	Restaurant Association NZ officially takes on the SeaStar rating system and increases number of staff assessing restaurants.	• Restaurants across New Zealand • Fisheries across New Zealand and Australia	Restaurant Association NZ investment	10 Years
CONNECTED — SCALE *Describe your plans for resilient scaling*	• Restaurants and retailers choose SeaStar certified sustainable seafood • Consumers preferentially consume seafood with the SeaStar certification • More fisheries choose to improve their practices and volunteer to be assessed to the SeaStar standard.	• Restaurants across the world • Fisheries across the world	• Restaurant Association NZ investment • Tourism NZ funding	15 Years

Fig. 10.8 Sustainable growth plan for the SeaStar case study. (Clara Schröder 2022)

2. Startup: Implement and refine the certification system through a real restaurant startup.

   The project remains local, testing in a real restaurant while gathering community feedback. It's also open to adjustments based on real-world challenges.

3. Autonomy: Establish financial sustainability by encouraging fisheries and restaurants to adopt the SeaStar standard.

   This stage ensures the project can self-sustain, transitioning from external funding to industry-driven incentives. The local engagement of restaurants and fisheries solidifies long-term adoption.
4. Replication: Expand the SeaStar certification by integrating it into the Restaurant Association of NZ.

   This stage supports open replication, ensuring governance shifts to a broader industry body while keeping alignment with original values. The decentralized approach allows independent adoption.
5. Scaling: Establish SeaStar as a widely recognized certification globally (for wider breadth of impact)

   The project is now connected internationally while maintaining local relevance by working with regional seafood industries. Scalability focuses on impact rather than just size, ensuring sustainable growth.

This is how a project like SeaStar can align with distributed, decentralized, and sustainable scaling, ensuring meaningful growth while maintaining its core mission of eco-conscious consumerism and marine restoration.

- Small: Begins with a student-led initiative to validate feasibility.
- Local: Prioritizes engagement with Auckland-based restaurants and fisheries.
- Open: Encourages industry participation and adapts based on community needs.
- Connected: Expands through partnerships with national and international organizations while preserving its mission.

Chapter 11
What Next? The Agile-to-Deep Iterative Process

Gabriela Nuri Baron

In the previous chapters, I have shown you many tools that can help teams understand, develop, and deliver innovative conservation projects. When deploying your project, it's time to move into a phase of continuous, adaptive improvement and sustained impact.

This is where the "Agile-to-Deep" Iterative Process becomes essential. Instead of seeing deployment as the endpoint, this approach encourages ongoing cycles of testing, refining, and impact measurement to ensure that the project not only lasts but thrives, bringing long-term benefits to the community (Fig. 11.1).

After launch, teams start with small improvements and establish feedback loops to gather immediate insights. Using low-cost, quick tools like the Impact Planning Canvas, teams can observe how well the project functions in its real-world setting. Ask questions like: Are the initial goals being met? Are we giving the community a voice? What adjustments could improve effectiveness? Gathering this feedback allows you to make small, responsive tweaks to better meet community needs. Each cycle of feedback adds new insights, deepening your understanding and strengthening the project's foundation.

In this phase, it's important to prioritize quality, resilience and meaningfulness over rapid growth. Follow nature's teachings and go slow, focusing on ensuring that the benefits are truly embedded in the community. Cultivate depth rather than breadth, refining processes, improving effectiveness, and aligning closely with the unique culture of the area. Remember that this approach values resilience and long-term success over immediate, large-scale impact.

Remember to keep *Reconnecting*, and always keep this as a guiding star. Nurture strong connections with local people, organizations, and ecosystems, aiming for

G. N. Baron (✉)
Faculty of Engineering and Design, University of Auckland – Waipapa Taumata Rau, Auckland, New Zealand
e-mail: g.baron@auckland.ac.nz

G. N. Baron (ed.), *Design for Environmental Conservation*,
https://doi.org/10.1007/978-981-96-9574-4_11

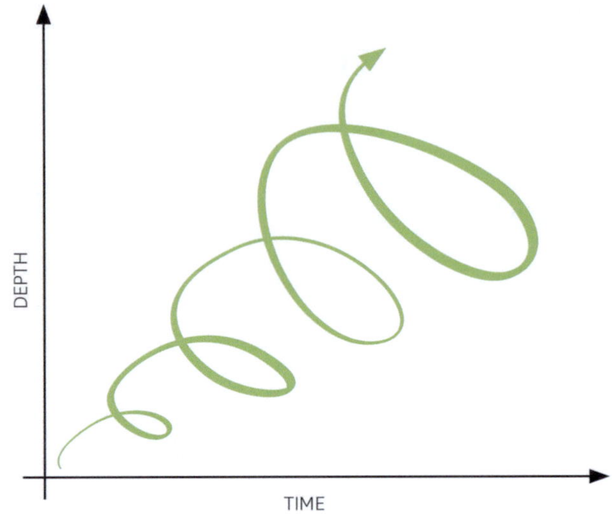

Fig. 11.1 The Agile to deep spiral, where projects are constantly refined using feedback loops. While starting agile, in terms of time and resources invested, the work is never finished, always reaching more depth in terms of understanding of the problem and efficacy/impact of the solutions in time

these relationships to become assets that remain long after the project has reached maturity.

Impact measurement can start with basic assessments and become more detailed as resources and knowledge grow. By building in this ongoing measurement, you increase credibility and gain valuable data to make adjustments, enhancing the project's resilience and capacity to attract funding.

As you gain confidence and secure additional resources, consider bringing in experts to deepen exploration of specific aspects needing improvement. Go back and refine the D4C Cnvases you had created, as these are meant to change and evolve as well. Customize tools as needed to address new challenges or evolving goals. You can even combine insights from different stages of the D4C methodology, tailoring the tools to suit the project's unique needs as it develops.

Above all, make sure the project's benefits—relationships, knowledge, revenue, and infrastructure—remain in the hearts, minds, pockets, and lands of the local community.

By building a project that adapts and grows with those it serves, you help create a legacy of empowerment, ensuring that the strengths, resources, and capabilities of the community grow alongside the project.

Chapter 12
Closing Thoughts

Gabriela Nuri Baron

Conservation efforts are often most needed in places where resources are scarce and communities are stretched thin. Grassroots organizations, in particular, face unique challenges in designing and managing conservation initiatives. D4C aims to bridge this gap by decentralizing conservation resources, making them accessible, adaptable, and inclusive. These tools are designed not to prescribe solutions but to inspire communities to create their own: solutions that honor their specific landscapes, cultures, and capacities.

As we bring this book to a close, I want to reflect on what we've explored together: a journey of reframing, relearning, and reconnecting with the natural world and each other. The Design for Conservation (D4C) methodology is not a rigid process; it is an invitation to embrace complexity with humility, creativity, and collaboration.

Gratitude and Hope

Working in conservation and sustainability often feels like an uphill battle. The magnitude of the challenges can be disheartening.

But even in moments of despair, there's something profoundly hopeful: we are not alone. Across the globe, there are countless people working toward the same goal: restoring and nurturing our planet, our Mother Earth. From citizens planting trees in their neighborhoods to scientists pushing the boundaries of knowledge, from Indigenous communities protecting ancestral lands to policymakers

G. N. Baron (✉)
Faculty of Engineering and Design, University of Auckland – Waipapa Taumata Rau, Auckland, New Zealand
e-mail: g.baron@auckland.ac.nz

G. N. Baron (ed.), *Design for Environmental Conservation*,
https://doi.org/10.1007/978-981-96-9574-4_12

advocating for systemic change, the collective force driving environmental restoration is immense and diverse.

Every small action adds to the larger momentum. Each conversation, project, and decision to care for the planet is part of a ripple effect that inspires others to act. Knowing that we are part of a global community—a growing movement of people dedicated to healing our Earth—helps transform the weight of the work into a shared responsibility. It reminds us that while single efforts have smaller impacts, together we are unstoppable: the dream of thriving planet is not just possible, it is already in motion. Ultimately, the hope behind this book is to inspire collaborative action. In navigating the challenges of environmental conservation, we've seen that no single discipline, community, or mindset holds all the answers. These challenges rarely have a clear start or finish, but they are opportunities for lasting partnerships and mutual growth. The D4C methodology is designed to foster these relationships, building networks of human and environmental care that extend beyond any single project's lifespan.

Online Resources and Tools

The Design for Conservation toolkit and all its resources are freely available at www.design4conservation.com. You can also access and download the tool collection via DOI here.

These tools are shared under a Creative Commons license, meaning they're open for anyone passionate about conservation to download, remix, adapt, and customize to suit their needs—just not for commercial purposes. If you use them, please cite the source to help keep knowledge flowing.

What You'll Find on the Design4Conservation Website:

- Downloadable Templates: All Design for Conservation templates are available as PDFs, ready for use. They're built to be flexible, so you can tweak them to fit your project's unique goals and community context.
- Video Tutorials: Step-by-step video guides walk you through the core tools in the toolkit, helping you integrate them seamlessly into your conservation work.
- Thriving Earth Meditation: A guided mindfulness exercise from the "Reconnect" phase, designed to deepen your connection with nature and your project's purpose.
- Newsletter: Sign up for the Design for Conservation newsletter to stay updated on new tools, case studies, and upcoming events.

- By keeping these resources free and accessible, Design for Conservation is here to help individuals and communities take conservation into their own hands—creating solutions that are locally rooted, globally connected, and built for lasting impact.

MIX
Papier aus verantwortungsvollen Quellen
Paper from responsible sources
FSC® C105338

If you have any concerns about our products,
you can contact us on
ProductSafety@springernature.com

In case Publisher is established outside the EU,
the EU authorized representative is:
Springer Nature Customer Service Center GmbH
Europaplatz 3, 69115 Heidelberg, Germany

Printed by Libri Plureos GmbH
in Hamburg, Germany